P9-CKJ-486

DESIGNING
CALIFORNIA
NATIVE
GARDENS

DESIGNING CALIFORNIA NATIVE GARDENS

The Plant Community Approach to Artful, Ecological Gardens

GLENN KEATOR *and*
ALRIE MIDDLEBROOK

A PHYLLIS M. FABER BOOK

UNIVERSITY OF CALIFORNIA PRESS

Berkeley Los Angeles London

University of California Press, one of the most distinguished university presses in the United States, enriches lives around the world by advancing scholarship in the humanities, social sciences, and natural sciences. Its activities are supported by the UC Press Foundation and by philanthropic contributions from individuals and institutions. For more information, visit www.ucpress.edu.

University of California Press
Berkeley and Los Angeles, California

University of California Press, Ltd.
London, England

Produced by Phyllis M. Faber Books
Mill Valley, California

Cover design by Nola Burger.
Interior design, photo editing, and typesetting by Beth Hansen-Winter.
Art coordination by Camille DaRocha.
Copy editing by Anne Canright.
Proof reading and index by Nora Harlow.

Unless otherwise noted, artwork is by Alrie Middlebrook and photography is by the authors. Contributing photographers: Saxon Holt, Stephen Ingram, Stephanie Morris, Greg Rubin, Rick Driemeyer, and April Owens. Selected illustrations by Ananda Yankellow. Our appreciation to the following landscape designers whose gardens are included in this book: Stephanie Morris, Greg Rubin, Rick Dreimeyer, Phil Johnson, Michael Thilgen, Paul Kephart, Kat Weiss, Anthony and Celia Ashley, Eric and Elisa Callow, Randi Swedenberg, Debbie Taylor, and many other talented designers.

Library of Congress Cataloging-in-Publication Data

Keator, Glenn.
    Designing California native gardens : the plant community approach to artful, ecological gardens / Glenn Keator and Alrie Middlebrook.
        p. cm.
    Includes bibliographical references and index.
    isbn-13: 978-0-520-23978-4 (cloth : alk. paper)
    isbn-13: 978-0-520-25110-6 (pbk. : alk. paper)
    1. Native plant gardening—California.   2. Native plants for cultivation—California.
I. Middlbrook, Alrie, 1944–   II. Title.

SB439.24.C2K43   2007
635.9'51794—dc22   2006051431

Manufactured in China

16   15   14   13   12   11   10   09   08   07
10   9   8   7   6   5   4   3   2

The paper used in this publication meets the minimum requirements of ansi/niso z 39.48-1992 (r 1997) (*Permanence of Paper*).

PAGE i: Western redbud flowers (*Cercis occidentalis*) against buckeye leaves (*Aesculus* sp.); PAGE ii: TOP TO BOTTOM: Red fescue (*Festuca rubra*) living roof in a coastal bluff environment. • Chaparral garden with no irrigation. • Oak woodland garden with grasses. • Matilija poppy (*Romneya coulteri*). • Creeping sage (*Salvia* 'Dara's Choice') in a mission style garden. Photograph by S. Morris. PAGE ii-iii: Mosaic of plant communities including chaparral, grassland, scattered oaks, and riparian woodland. PAGE v: Fawn lily (*Erythronium klamathense*).

The publisher gratefully acknowledges the
generous contribution to this book provided by
the Stanley Smith Horticultural Trust.

CONTENTS

CONTENTS

FOREWORD

The authors, Glenn Keator, a botanist and teacher, and Alrie Middlebrook, an artist, garden designer, and owner of Middlebrook Gardens in San Jose, met on a field trip Glenn was leading to the White Mountains, home of the ancient bristlecone pine. Glenn fell in love with California's native wildflowers during his childhood years of camping in the Sierra Nevada. Since those early times, he has experimented with growing natives in his own gardens as well as teaching hundreds of people how to identify and appreciate native species of plants and trees. He has a great following of people who have become ardent native plant aficionados.

As Glenn pointed out the flowers and shrubs growing among the bristlecones, Alrie became enchanted with the beauty and diversity of this natural garden in the mountains. The gardens she had designed up until that moment had been planted with an indiscriminate mix of subtropical flowers, Mediterranean plants, and a few natives. She felt she was ignorant of the thousands of species that make up the palette of California. Alrie says she knew on that trip that she had struck the mother lode. Since then she has traveled up and down the state observing and increasing her understanding of our native plant communities, with one idea in mind: to create gardens using our spectacular native flora.

Together Alrie and Glenn decided to collaborate and combine Glenn's botanical knowledge and Alrie's artistic sense and horticultural skills to encourage Californians to "go native" and recreate the state's natural beauty. They jointly taught a series of classes at the San Francisco Botanical Garden's Strybing Arboretum, learning as they went what was entailed in building a native garden and visiting the few extant native gardens. During this period, Alrie was making the transition in her company to create native-style gardens, and they were able to use these gardens in their teaching.

This book has emerged from the series of classes Glenn and Alrie have taught over the past decade. All but three of the chapters deal with specific gardens that Alrie has built; the other three represent her conceptual designs. She tries to take advantage of the natural setting of a garden site, what grew there before the site was developed, to arrive at plants that are best adapted to the local soils and climate. Alrie says her greatest pleasure is recreating features in gardens that occur in nature, like a sunny wildflower meadow at the edge of a pond; a shady, winding woodland path; or a steep bluff with rocky outcroppings. The goal is both an ecologically sound garden, and a beautiful one as well.

Alrie has taken a somewhat unconventional approach with her plan view representations, because she feels the home gardener needs a more accessible image than

Santa Cruz Island ironwood (*Lyonothamnus floribundus* ssp. *aspleniifolius*), Santa Barbara Botanic Garden. Photograph by S. Ingram.

the usual architectural drawings that landscape architects work with. She has therefore created hybridized plan/elevation drawings to make it easier to visualize the finished garden. She has also on occasion included two-dimensional landscape drawings as backdrops for the gardens, to suggest the natural setting that is so beautiful and so integral to the overall design.

Transferring concepts to a workable book that conveys the awesome diversity of California's native communities, that describes the attributes of the many ornamental species, and that translates natural landscapes into workable gardens has not been an easy task for Glenn and Alrie. This project is the fruit of many discussions and field visits. And when all of the planning considerations have been addressed, one more challenge remains: which species to recommend? Of the two thousand or more possibilities, only a relative few are readily available from commercial sources. For this reason, they have constantly reevaluated the palette of plants they present in this book. Their latest roster of species still suffers to a degree from problems of availability; certain communities such as montane meadows are challenging because few nurseries are propagating the plants that grow there. In some cases choices were driven by appropriateness to habitat as well as sheer beauty.

Alrie and Glenn hope California gardeners will be inspired and encouraged to take their own steps to preserve our rich natural heritage by creating native gardens imitative of the surrounding natural ecosystems wherever they happen to garden. These gardens use less water, fertilizers, and soil amendments, and they attract and support local wildlife, most notably hummingbirds and native bees. But just as important, they engage our senses and help reconnect us with our own local landscape and beauty as nature defines it.

Phyllis M. Faber

CALIFORNIA'S NATIVE PLANT COMMUNITIES

An Overview

by Glenn Keator

We Californians are fortunate to live in a mild Mediterranean climate where we can grow an amazingly broad palette of plants from many places around the world.

So why should we turn to California natives in preference to these others? The most compelling reason is to create a sense of place. Today, native plant communities surround our urbanized areas; indeed, our homes are built on land that once was covered by these plants. What better way is there to remind ourselves of this special geographic region we call home than to recreate, in our own yards, the native gardens found in the wild? Anyone can have a garden with roses (mostly hybrids from China and Europe), petunias (from South America), fuchsias (from mountainous South and Central America), and impatiens (many from Africa). But natives tell about where we live; they make us feel at home.

Plant communities have evolved over thousands of years, adapting to the changing climate and to local soil conditions. What we see today is the result of competitive forces between plant species and between plants and animals, particularly plant pollinators. Why not take advantage of nature's work and recreate these relationships in our own gardens? Preserving the integrity of the local biotic community can give endless rewards to us and to the wildlife we may have displaced.

Here are several good reasons for establishing a native garden on your site:

1. Local native plants are already well adapted to the conditions of climate and soil, making them easy to grow and likely to succeed. You can wander farther afield, into other native communities of plants whose basic requirements are similar to those of your specific place, to increase your diversity of species.
2. Although there is no such thing as a maintenance-free garden, using plants well adapted to the site translates into low garden maintenance: less watering, fertilizing, and few or no soil amendments.
3. Native plants attract native pollinators. Although many of our pollinating bees and butterflies are generalists, native pollinators are often extremely narrow in their preferences and cannot survive without native flowers to feed on or lay their eggs. Providing appropriate plants for these pollinators helps both plants and wildlife.

4. By growing native plants in your garden, you can save money and water. This is especially true if you eliminate lawns, which require copious amounts of water, fertilizers, and herbicides, to say nothing of the labor involved in maintenance. Most natives, once established, require little additional summer water and no amendments.

5. Using natives eliminates the need for toxic substances in the garden—herbicides, artificial fertilizers, and pesticides—that create harmful runoff or otherwise pollute the environment.

6. Native plants do not disturb the site as much as many exotic plants do. Their installation is simple, and soils do not require constant rototilling and amending.

7. Native plants are highly diverse and offer species for any garden site or situation. Natives are not restricted to a single watering regime, kind of soil, or areas of limited or special microclimates. Natives run the gamut from wet, foggy, coastal bluffs to arid, hot deserts and from mild, frost-free winters to prolonged subzero winters with a heavy snow cover. Once established, they carry on as they do in the wild with little or no maintenance.

8. Natives are aesthetically beautiful. Some of the longest-blooming, showiest

Desert landscape with wildflowers.

flowers are produced by native species. Some of the best foundation shrubs and trees are native. And some of the most beautiful foliage plants for texture, color, and fragrance come from native species. (Although not all natives are attractive, hundreds of species have garden potential.) Like other plants from other parts of the world, natives need to be carefully selected for the right place in the garden. Knowing when to plant, when to cut back, how to prune, and where to use a given species is crucial to the success of its garden appeal.

Most of the negative press that natives have received is due to misconceptions about how to use, care for, and install them. Like other garden plants, choosing the right species requires attention to details, knowledge of growth patterns, and cultural requirements.

There is no doubt that gardens featuring other Mediterranean-climate plants also work well in California gardens with minimal care, watering, and fertilizing as compared to the more commonly used exotic plants. Many of these plants are well adapted to our climate, are highly attractive, and thrive in foothill and coastal areas. But the point of this book is to make a case for going native, both to create gardens of beauty and to give a sense of place and extend native habitats into urban settings.

The other side of the nonnative "coin" is that some nonnative species grow so well they have "escaped" from gardens to become aggressive and invasive weeds in native landscapes. The reason for the success of these plants, which may sometimes grow better than native species, is a lack of their homeland pests and diseases, which otherwise keep them in balance.

Finally, while many Mediterranean natives look attractive growing with our own natives, many South African and Australian species do not blend harmoniously with California natives because of their different growth patterns and floral designs. Such plants should be used by themselves.

The goal of our book is thus threefold:

1. To acquaint gardeners with wild places and plant communities that have occurred where they live. Today, most urban and suburban areas have, at best, the fragmented remains of these original wild places. Familiarity with our natural heritage promotes a sense of place and a desire to reconnect to nature.
2. To describe a wide palette of species from each plant community with some of their maintenance requirements and means of propagation. Each chapter features 25 or more species that can be used to create a garden based on a single native plant community. Additional species, often less common in the trade or poorly known, are suggested as well, to flush out the roster of possibilities.

3. To provide sample plans of native gardens based on specific plant communities. In most cases, examples of existing gardens include plant lists and planting plans. Many of these gardens feature more than one plant community because of differences in slope, shade, moisture availability, and soil texture.

This book is organized into three sections:

- Introductory material describes the philosophy behind the book, the organization of the book, and gives background on plant communities. Two transects across the state illustrate how plant communities relate to California's geography. In addition, the principles of a "design ethic" for California are outlined in full.
- The twelve community-based chapters—the heart of the book—aim toward a new gardening paradigm: creating native gardens that are aesthetically beautiful and ecologically sound.
- Appendices give sources of natives, relevant books and websites, seasonal gardening tasks, and a glossary.

About Community-Based Gardens

Besides staying true to what nature intended for your special place, the astute gardener carefully chooses hardscape, art objects and other amenities, and methods of construction that tread lightly on the earth and complement the natives chosen. Such designs blend well with the site. California natives lend themselves to a wide array of different sites, design considerations, and styles. For example, a chaparral garden needn't look out of place on a property with a formal house. The right placement of material and the careful use of cultural amenities can achieve beauty and natural balance.

The twelve plant community chapters have been selected to give most California gardeners examples they can use wherever they live. We've tried to choose the most representative plant communities from the coast, foothills, valleys, and middle elevations throughout the state. Here are the twelve selections:

Bluffs and cliffs: elements for a rock garden. Rock garden enthusiasts need look no farther than coastal bluffs or inland rocky outcrops for a source of appropriate material. Alpine gems from the mountains of Europe and China are not required to satisfy the desire for beauty in these special landscapes.

Redwood forest: gardening under cool giants. Even though many gardens don't have a redwood tree, other conifers that grow in coastal climates serve equally well. Spring wildflowers, seasonal shrubs, ferns, and pleasing ground covers round out a palette for a redwood forest design.

Coastal sage scrub: southern California's "soft" chaparral. Coastal sage scrub is a maligned and endangered community; great swaths have been lost to development. Yet this community contains shrubs and perennials with great ornamental promise when properly grown. Creating a coastal sage garden helps reverse habitat loss.

The Channel Islands: a parade of unique plants. The mild year-round climate of the Channel Islands off southern California harbors special plants, including several found nowhere else, and lends a refinement to coastal gardens. Lacy ironwoods and rose-flowered buckwheats vie for a place in the sun.

Deserts: juxtaposing plants from an extreme habitat. Even though relatively few people live in the true deserts of southern California, many live in peripheral areas where there is scant winter rainfall and ample summer sun and heat. The desert is an inspirational source of plant material, providing a plethora of colorful shrubs, subtropical trees, spiky cacti, and vividly colored wildflowers.

Montane meadows: gardening with mountain wildflowers. Who hasn't had the magical experience of walking through a colorful summer meadow in the high mountains? Many

Montane meadow plant community near Lake Winnemucca.

gardens in the higher foothills are perfect places to recreate such unrivaled beauty. Experimentation in the lower foothills may also lead to designs that mimic these places of the imagination.

Mixed-evergreen forest: summer shade between fog and sun. Many places in California experience summertime conditions that combine the dry heat of interior valleys and the foggy cool of the coast. Such sites are perfect for a mixed-evergreen forest, blending large evergreen trees such as Douglas-fir and madrone with a varied understory of shrubs, perennials, and ferns.

Oak woodland: California's signature foothill landscape. Many fortunate gardeners already have oaks on their property, yet many ornamentals require the summer water that slowly kills these magnificent trees. California's oak woodlands provide a fine palette of plants perfectly adapted to grow under oaks.

Grasslands: paradise for wildflowers. Grasslands can be planted on their own as alternatives to lawns or combined with a wide variety of colorful wildflowers that bloom from early spring into summer. Garden grasslands can be watered to remain green most of the year, or left to go dormant in the summer as they do in nature.

Fawn lily (*Erythronium helenae*) with Indian warrior (*Pedicularis densiflorus*).

Chaparral: drought-adapted shrubs for the garden. Water-conserving gardens that receive abundant summer sun are perfect places to feature this plant community. Chaparral shrubs need minimal care once established and often have dramatically beautiful leaves and flowers.

Riparian woodlands: a plant palette for heavy soils. Gardens with their own stream or other permanent water feature lend themselves to plants from permanent watercourses. Such plants are also appropriate where soils are heavy, drain poorly, or lie next to a part of the garden that receives regular summer water.

Wetlands: the beauty of water in the garden. Besides the riparian woodlands described above, even the driest gardens benefit from water features such as bird baths, wine barrels filled with water, or ponds. The juxtaposition of drylands and water is ageless in garden design and provides a hint of lushness in the starkest landscapes.

Plant Communities in California

Plant communities are repeatable assemblages of plants that grow together because of similar adaptations to microclimates, soils and slopes, and biotic factors. Examples include redwood forest dominated by coast redwood (*Sequoia sempervirens*), freshwater marsh dominated by grasslike plants such as tules (*Scirpus* spp.) and cattails (*Typha* spp.), and chaparral dominated by several different evergreen, drought-tolerant shrubs such as wild lilacs (*Ceanothus* spp.) and manzanitas (*Arctostaphylos* spp.).

Climates are defined by the interaction of precipitation, temperature, wind, and fire. The variations in the cycles of rain- and snowfall together with the daily and seasonal fluctuations in temperature are the main determinants of climate.

Soils and slopes both play pivotal roles in determining vegetation. North- and east-facing slopes receive more shade in summer and remain cooler and moister, whereas south- and west-facing slopes receive considerable afternoon sun, resulting in hot, dry conditions.

Soils vary greatly in their depth, fertility, texture, and pH, all of which help determine which species survive and thrive. California's diverse rocks, acted on by time, climate, and the plants growing on them, lead to the soil type found at any one site. Coarse, sandy soils are well oxygenated, poor at retaining water, and low in mineral nutrients. Clay soils are poorly oxygenated, retain water well, and are generally high in nutrients. The pH measure indicates the alkalinity or acidity of soils. Most plants thrive best in a nearly neutral pH (7.0), but some prefer acid soils (pH 4–6.5) and a very few do best in alkaline soils (pH 7.5–8.5).

Biotic factors, which include all the interactions between the varied organisms—

fungi, bacteria, animals, and plants—that live in a given community, are complex and remain poorly understood. Examples include the kinds of pollinators available to service flowers and the types of soil fungi that inhibit or promote the growth of plant roots.

California is among the world's most diverse places geologically and botanically. Varied topography results in myriad microclimates. Foothill California is dominated by a Mediterranean climate—a pleasing mix of mild, wet winters; rainy springs; and hot, dry summers and falls. Many native plants are beautifully adapted to this mix. Yet there are many variations: the coastal fog belt experiences cool, humid summers where fog condenses on tree branches to drip to the ground as summer "rain." The high mountains have a more continental climate, with harsh, cold, snowy winters; sudden springs; and short, warm summers punctuated by occasional thunderstorms. The deserts embrace a combination of sere conditions with occasional precipitation (usually less than ten inches annually) and long periods of excessively hot temperatures.

Typical Mediterranean-climate communities include seasonal grasslands, chaparral, coastal sage scrub, and oak and mixed-evergreen woodlands. Characteristic coastal fog-belt communities encompass coastal scrub, redwood forest, closed-cone pine/cypress forest, coastal dunes, and coastal bluffs. Mountain climates foster mixed conifer-

Grassland meadow with goldfields (*Lasthenia glabrata*) and poppies (*Eschscholzia* ssp.).

ous forest, montane chaparral, montane meadows, aspen forests, subalpine forest, and alpine fellfields. Desert climates embrace pinyon juniper woodland, sagebrush scrub, creosote bush scrub, desert washes, desert oases, and cactus-succulent scrub.

Soils derived from serpentine and limestone rocks favor a unique set of species resulting in a different version of local communities: serpentine-influenced chaparral and grasslands look different from ordinary chaparral and grasslands.

Communities with a dependable year-round source of water are also very different. Riparian woodlands and forests line the flood plains of rivers and permanent streams; lakes and marshes are home to suites of plants adapted to freshwater or salt marsh conditions; ponds and small lakes have their own assemblage of species. Such specialized wetlands as bogs are dominated by sphagnum moss and are home to unusual ferns, members of the heather family (Ericaceae), orchids, and insectivorous plants.

Slope plays a pivotal role in determining which communities occur in a given area. While a south slope might support chaparral, a comparable north slope would be home to foothill or oak woodland. Where a south slope wears a mantle of oak woodland, a corresponding north slope finds mixed-evergreen or redwood forest.

In summary, plant communities are determined by a combination of climate (including local microclimates), soils, slope, and subtle biotic factors. Often, two, three, or even four different communities can occur next to one another because of differences in slopes, soils, and availability of water. A common example is where canyon bottoms meet adjacent slopes. The canyon may have a permanent stream supporting riparian woodland, while north-facing slopes are covered with oak woodland, south-facing slopes with chaparral, and serpentine outcrops sustain a modified conifer woodland, grassland, or stunted chaparral.

Plant Community Transects

Two transects across the state are described below. The first heads east from the San Francisco Bay region, and the second starts in the Channel Islands near Ventura and likewise traverses east. These transects show the relationship between California's complex geology, terrain, climates, and plant communities, demonstrating how physical geography influences plant growth and distribution.

Not all of the communities and species described here are included in later chapters of this book; for example, few salt marsh or alpine plants adapt well to gardens.

Transect across Central California

By the edge of the sea, slope and soils have a large influence. Steep coastal bluffs have matted and cushion-forming plants, including coast rockcress (*Arabis blepharophylla*), Menzies and Franciscan wallflowers (*Erysimum menziesii* and *E. franciscanum*), varicolored lupine (*Lupinus variicolor*), coast paintbrush (*Castilleja wightii*), California phacelia (*Phacelia californica*), brownie thistle (*Cirsium quercetorum*), lizard-tail (*Eriophyllum stachaedifolium*), and bluff chickweed (*Cerastium arvense*). Gentler slopes are covered with north coastal scrub, a mixture of small to medium-sized evergreen and semievergreen shrubs such as coyote brush (*Baccharis pilularis*), coffeeberry (*Rhamnus californica*), bush monkeyflower (*Mimulus aurantiacus*), California sagebrush (*Artemisia californica*), blue-witch (*Solanum umbelliferum*), blueblossom ceanothus (*Ceanothus thyrsiflorus*), and silk tassel bush (*Garrya elliptica*).

The deepest soils on blufftops support a grassland called coastal prairie. Coastal prairies are home to native bunchgrasses such as red fescue (*Festuca rubra*), Nutka reed grass (*Calamagrostis nutkaensis*), and tufted hair grass (*Deschampsia caespitosa*), and a colorful assortment of bulbs, perennials, and annual wildflowers. Among the floral pageant here are Douglas iris (*Iris douglasiana*), blue-eyed grass (*Sisyrinchium bellum*), yarrow (*Achillea millefolium*), checkerbloom (*Sidalcea malviflora*), goldfields (*Lasthenia* spp.), checker lily (*Fritillaria affinis*), California poppy (*Eschscholzia californica*), dwarf brodiaea (*Brodiaea terrestris*), and baby-blue-eyes (*Nemophila menziesii*).

The poorest soils on the foggiest promontories support closed-cone pine/cypress forest, a community with even-aged stands of Monterey, bishop, and beach pines (*Pinus radiata*, *P. muricata*, and *P. contorta* ssp. *contorta*) sometimes in company with Monterey, Santa Cruz, and pygmy cypresses (*Cupressus macrocarpa*, *C. abramsiana*, and *C. goveniana* ssp. *pygmaea*). The plants that live under the canopies of these trees have much in common with the forest floor of other foggy forests, such as redwood forest and—farther north—north coastal coniferous forest.

Stream and river mouths often meander into freshwater and, where the tides come in, coastal salt marshes. Salt marshes are important places for birds and aquatic animal life but are poor in plant species and of little value to the average gardener. Marshes, however, are often bordered by beaches and extensive sand dunes that are home to interesting plants of the coastal strand. Plants here form dense cushions or low, sprawling mats anchored by extensive, often fleshy roots. Many colorful perennials abide these conditions, including sand-verbenas (*Abronia* spp.), beach suncups (*Camissonia cheiranthifolia*), various lupines (*Lupinus* spp.), beach morning glory (*Calystegia soldanella*), dune sagebrush (*Artemisia pycnocephala*), beach bursage (*Ambrosia chamissonis*), beach strawberry (*Fragaria chiloensis*), beach sweet pea (*Lathyrus littoralis*), and dune gumplant (*Grindelia stricta* var. *platyphylla*).

Lying a half-mile back from the ocean's edge are deep alluvial flood plains and slopes covered with redwood forest. Redwood forest thrives best on deep, silty soils within the fog belt, increasing in vigor and in the lushness of its understory northward to the Oregon border, where winter rainfall is high. Redwood forests are home to many berry-producing shrubs, rhododendrons, and other heather relatives, ferns, redwood sorrel

Mixed-evergreen forest with California bay (*Umbellularia californica*) at Huckleberry Preserve in the Oakland hills.

(*Oxalis oregana*), wild ginger (*Asarum caudatum*), and shade-loving members of the lily family such as trilliums (*Trillium* spp.), bead-lily (*Clintonia andrewsiana*), false lily-of-the-valley (*Maianthemum dilatatum*), fairy bells (*Disporum* spp.), and slink pods (*Scoliopus bigelovii*).

On steep slopes or where logging has taken its toll, redwood forest is diminished. Under these circumstances, redwoods are often mixed with other evergreen coniferous and hardwood trees. Eventually this mixed-evergreen forest replaces redwood forest on the drier upland slopes, or where summer fog has less influence. The mixture of trees varies from place to place according to locale and fire history. It includes madrone (*Arbutus menziesii*), tanbark oak (*Lithocarpus densiflorus*), coast live oak (*Quercus agrifolia*), coast chinquapin (*Chrysolepis chrysophylla*), California nutmeg (*Torreya californica*), California bay (*Umbellularia californica*), and Douglas-fir (*Pseudotsuga menziesii*).

As you move inland away from the summer fog's influence or onto high slopes above the fog, the plant communities change again. The hottest, driest slopes are covered with "hard" chaparral—a dense mixture of evergreen shrubs with stiff, drought-resistant leaves. Chaparral is home to some of our finest colorful, drought-tolerant shrubs, including many wild lilacs (*Ceanothus* spp.), manzanitas (*Arctostaphylos* spp.), bush poppy (*Dendromecon rigida*), coffeeberry (*Rhamnus californica*), fremontia (*Fremontodendron* spp.), silk tassel (*Garrya* spp.), chaparral pea (*Pickeringia montana*), yerba santa (*Eriodictyon* spp.), mountain mahogany (*Cercocarpus betuloides*), scrub oaks (*Quercus berberidifolia* and *Q. durata*), and chamise (*Adenostoma fasciculatum*).

Other scattered communities on hot, dry slopes are inland versions of closed-cone pine and cypress forests. Restricted to nutrient-poor serpentine and sandstone soils, these forests display even-aged stands of knobcone pine (*Pinus attenuata*) or dense, dwarf woodlands of Macnab and/or Sargent's cypresses (*Cupressus macnabiana* and *C. sargentii*). Like the chaparral, these trees are adapted to proliferate after natural wildfires that sweep through every twenty to forty years.

Gently rolling slopes are favored by oak woodlands or foothill woodland—a mixture of oaks (*Quercus* spp.), California buckeye (*Aesculus californica*), and gray pine (*Pinus sabiniana*). This landscape portrays the essence of California and is home to many fine shrubs, bunchgrasses, and perennial wildflowers. Valley bottoms have grassland meadows. Grasslands also extend onto gently rolling hills and even steep slopes, where grazing pressure has eliminated shrubs and trees. Our grasslands are the finest repositories for colorful wildflowers and bulbs. Even today they display a breathtaking beauty in years of abundant rainfall. Just a few examples of the wildflowers include bulbs such as mariposa-tulips (*Calochortus* spp.) and brodiaeas (*Brodiaea*, *Dichelostemma*, and *Triteleia* spp.), and annuals such as phacelias (*Phacelia* spp.), goldfields (*Lasthenia* spp.), glueseed (*Blennosperma nanum*), redmaids (*Calandrinia ciliata*),

creamcups (*Platystemon californicus*), poppies (*Eschscholzia* spp.), gilias (*Gilia* and *Linanthus* spp.), clarkias (*Clarkia* spp.), lupines (*Lupinus* spp.), tarplants (*Madia* spp.), and tidy-tips (*Layia* spp.).

All through this region, from the immediate coast across the Coast Ranges to the Central Valley, watercourses are accompanied by narrow corridors of riparian woodlands whose composition changes according to how cool or hot the summers are. Riparian woodlands are dominated by such fast-growing, deciduous trees as ashes (*Fraxinus* spp.), black walnut (*Juglans californica*), maples (*Acer* spp.), western sycamore (*Platanus racemosa*), cottonwoods (*Populus* spp.), and alders (*Alnus* spp.). Smaller trees such as willows (*Salix* spp.) and elderberries (*Sambucus* spp.) also abound. The trees are often festooned with such exuberant vines as wild grape (*Vitis californica*), clematis (*Clematis ligusticifolia*), and vine honeysuckles (*Lonicera* spp.).

Many of the patterns we see today have been altered by suburban sprawl, extensive grazing, and field and fruit agriculture, in addition to the introduction of aggressive, nonnative weeds. Invasive species that outcompete native species have reduced plant diversity markedly. Most grasslands currently consist of alien grasses—wild oats (*Avena* spp.), Italian rye grass (*Lolium perenne*), ripgut brome (*Bromus diandrus*), foxtails (*Hordeum* spp.), and many more. Besides native wildflowers, there are also such invasive "wild" flowers as chicory (*Cichorium intybus*), milk thistle (*Silybum marianum*),

Oak trees and boulders, Morgan Territory Regional Park, Contra Costa Co.

bull thistle (*Cirsium vulgare*), yellow star thistle (*Centaurea solstitialis*), poison hemlock (*Conium maculatum*), and Queen Anne's lace (*Daucus carota*).

The Central Valley was once home to a vast, convoluted quiltwork of marshes, riparian forests, valley oak savannah, and valley grassland. Most of these communities have either been obliterated or left as fragments because of agriculture and housing developments. On the far side of the valley, the Sierra Nevada begins with a series of gently folded foothills that gradually give way to steep, mid-elevation mountains and canyons, terminating in high, rugged, picturesque peaks. Because the rise from the foothills to the mountain crest is gradual, there are many different plant communities at different elevations. You can accurately gauge where you are by looking at the plants around you.

The foothills start with the same mix of communities as occur in the inner Coast Ranges: oak and foothill woodlands, grasslands, and hard chaparral. Somewhere between 2,000 and 3,000 feet there is a transition to the yellow pine or mixed coniferous forest, which creates a broad belt extending up another three to four thousand feet. This area receives thirty to forty inches or more of annual precipitation and supports some truly large specimens of conifers, including ponderosa and sugar pines (*Pinus ponderosa* and *P. lambertiana*), incense-cedar (*Calocedrus decurrens*), white fir and Douglas-fir (*Abies concolor* and *Pseudotsuga menziesii*), and—in scattered, well-watered groves—giant sequoia (*Sequoiadendron giganteum*). Understory broadleaf trees include bigleaf maple (*Acer macrophyllum*), mountain dogwood (*Cornus nuttallii*), California bay

Ribbon Falls and yellow pine (*Pinus ponderosa*) forest, Yosemite Valley.

(*Umbellularia californica*), and black and goldcup oaks (*Quercus kelloggii* and *Q. chrysolepis*). Wet areas along permanent streams are home to temporary meadows filled with wildflowers, while steep rocky slopes or recent burns are covered with a montane chaparral of bitter cherry (*Prunus emarginata*), greenleaf manzanita (*Arctostaphylos patula*), snowbrush ceanothus (*Ceanothus cordulatus*), huckleberry oak (*Quercus vaccinifolia*), and mountain chinquapin (*Chrysolepis sempervirens*). Riparian woodlands still accompany stream courses.

Above this broad middle-elevation zone—somewhere between 6,000 and 7,500 feet—the more austere subalpine forests begin. These forests vary according to fire regime, soils, and the severity of winters. Many areas are covered with monotonous stands of lodgepole pine (*Pinus contorta* ssp. *murrayana*), Jeffrey pine (*P. jeffreyi*), or red fir (*Abies magnifica*), but the highest bits of forest have an open canopy of red fir, lodgepole pine, western white pine (*P. monticola*), mountain hemlock (*Tsuga mertensiana*), Sierra juniper (*Juniperus occidentalis*), and limber pine (*P. flexilis*). These species are never all present at one place. At the very highest reaches of tree growth—just below timberline—the beautifully wind- and snow-sculpted whitebark pine (*P. albicaulis*) dominates. In this same zone, wet valley bottoms and old lakes that have silted up are home

Bristlecone pine (*Pinus longaeva*) in the White Mountains.

to meadows; steep rocky slopes have montane chaparral; and meadows are fringed with quaking aspen (*Populus tremuloides*).

Mountain meadows from mid-elevation to above timberline are home to magnificent summer perennial wildflower displays and provide a rich source of beautiful material for mountain gardens. Besides a winter-dormant framework of varied grasses and sedges, the wide array of perennials and bulbs includes corn-lily (*Veratrum californicum*), monkshood (*Aconitum columbianum*), red columbine (*Aquilegia formosa*), lilies (*Lilium* spp.), camas (*Camassia quamash*), mountain iris (*Iris missouriensis*), shooting stars (*Dodecatheon* spp.), yampahs (*Perideridia* spp.), green gentian (*Swertia radiata*), gentians (*Gentiana* spp.), monkeyflowers (*Mimulus* spp.), cinquefoils (*Potentilla* spp.), paintbrushes (*Castilleja* spp.), elephant snouts (*Pedicularis groenlandica*), and many more.

In the very highest places—from 8,500 feet in the extreme north to 12,000 feet in the south—timberline is reached. Above timberline, the growing season is too short, the winds too severe, and the snows too heavy to allow tree growth. Those trees that manage to survive are stunted shrubs referred to as krummholz. The alpine zone is home to matted cushion plants among rocks—the so-called alpine fellfields—or short tussocks of grasses, sedges, and wildflowers in alpine meadows. Many beautiful wildflowers are found in this zone, but only a few readily adapt to lowland gardens.

Beyond the high Sierran crest, the land falls steeply and quickly to the high desert beyond, often within a span of only a few miles. This side of the great mountains is dry because it lies in the rain shadow of the mountain crest. Vegetation here is influenced by drought. Subalpine forest spills over a short ways, then as you descend is replaced by Jeffrey pine, lodgepole pine, and white fir. Stream courses are accompanied by quaking aspen and, to the south, water birch (*Betula occidentalis*). Meadows occur under the right conditions, and rocky slopes wear a mantle of montane chaparral.

At the base of the Sierra, the arid, cold desert begins. This vast high desert stretches across Nevada and Utah all the way east to the Rocky Mountains. It is dominated by open woodlands of junipers (*Juniperus* spp.) and pinyon pines (*Pinus monophylla* and others) intertwined with large swaths of sagebrush scrub. The fragrant sagebrush scrub is home to big sagebrush (*Artemisia tridentata*), rabbitbrush (*Chrysothamnus* spp.), antelope brush (*Purshia tridentata*), desert peach (*Prunus andersonii*), hopsage (*Grayia spinosa*), and Mormon tea (*Ephedra* spp.).

Southern California Transect

Our southern transect starts near Ventura. The four northern Channel Islands, located ten to fifty miles off the coast, are a prominent feature here. They are evolutionary laboratories because of their isolation and gentle climates, and several endemic plants—some relicts from the past, and others newly evolved—make their home here.

Santa Cruz Island, whose nearly one hundred square miles make it the largest island, displays an impressive mix of plant communities. Few of these communities differ greatly in appearance from ones of the mainland despite their different species compositions, but the ironwood groves deserve special mention. The island ironwood (*Lyonothamnus floribundus*) creates tight-knit copses in moist canyons and arroyos. Typical understory plants include island cherry (*Prunus ilicifolia* var. *lyonii*), toyon (*Heteromeles arbutifolia*), island scrub oak (*Quercus pacifica*), island paintbrush (*Castilleja lanata* var. *hololeuca*), white globe-tulip (*Calochortus albus*), island bush monkeyflower (*Mimulus flemingii*), and southern Humboldt lily (*Lilium humboldtii* var. *ocellatum*).

Other island plant communities include closed-cone pine forest, oak woodland, hard chaparral, coastal sage scrub, riparian woodland, grassland, coastal strand, and coastal bluff. Of these, only coastal sage scrub is new to our discussion.

On the mainland, coastal sage scrub is a threatened plant community because of extensive urbanization along the coast and in the coastal mountains. This southern counterpart to north coastal scrub lives on rocky slopes along the coast. It may also occur inland in areas that receive summer fog or that have nutrient-poor soils. Coastal sage scrub (also known as "soft" chaparral) consists of low, fragrant shrubs—many of them sage scented—such as California sagebrush (*Artemisia californica*), black sage

California sagebrush (*Artemisia californica*) at Point Lobos State Reserve.

(*Salvia mellifera*), purple sage (*S. leucophylla*), and white sage (*S. apiana*). Other prominent shrubs include lemonade berry (*Rhus integrifolia*), laurel sumac (*Malosma laurina*), chaparral yucca (*Hesperoyucca whipplei*), California buckwheat (*Eriogonum fasciculatum*), and in some places prickly pear and cholla cacti (*Opuntia* spp.). It commingles inland with hard chaparral, which is prominent on the steep mountain slopes of southern California.

Missing along the mainland coast are closed-cone pine forest (a Torrey pine forest can be found near San Diego, and a limited forest of bishop pines near Lompoc) and redwood forest, but coastal prairie, coastal strand, and marshes are present in the same patterns as farther north. Watercourses are lined with riparian woodlands. Prominent in these woodlands are western sycamore, white alder, and California black walnut fringed by coast live oak and California bay.

The main plant communities encountered inland on the rugged mountains are patchworks of hard chaparral, coastal sage scrub, oak woodland, and grassland, but there are only hints of mixed-evergreen forest in the most protected canyons or on north-facing slopes. Many southern California mountains are surprisingly rugged and steep, with several surpassing 8,000 feet in elevation. For example, Mt. Pinos in the northeastern corner of Ventura County is high enough to support Sierran communities such as ponderosa pine forest, montane meadow, montane chaparral, and subalpine forest. The elevations at which these communities occur here is considerably higher than at the more northerly locations; the farther south you go, the higher the elevations of the corresponding plant communities, because precipitation and winter temperatures change with latitude. There is also an admixture of desert elements in these mountain communities.

The east side of the mountains lies in a rain shadow with a corresponding diminution of precipitation. Chaparral and oak woodland may quite abruptly give way to a true desert flora. The vast Mojave Desert lies east of the last foothills. The Mojave is an intermediate desert lying at elevations higher than those of the Sonoran Desert to the south but lower than the cold desert east of the Sierra. The term *intermediate*, however, is misleading, for the Mojave is not a vast plain but a whole series of plains, basins, and mountain ranges.

Mojave desert vegetation goes like this: the plains and basins are covered with creosote bush scrub and shadscale scrub. Shadscale scrub is adapted to salty soils and rims dry salt-lakes and low basins where salts accumulate. It is dominated by short, scruffy shrubs such as budsage sagebrush (*Artemisia spinescens*), shadscale saltbush (*Atriplex confertifolia*), and other saltbush relatives. Creosote bush scrub is on better-drained or less salty soils, and covers thousands of acres with widely spaced creosote bushes (*Larrea tridentata*) interspersed with burro bush (*Ambrosia dumosa*) and, sometimes, a mixture of other colorful shrubs. When rains have come at precisely the right time and in the right amount—on the order of every ten to thirty years—the scrub is illuminated by waves of colorful annual wildflowers: desert dandelion (*Malacothrix glabrata*), thistle sage (*Salvia carduacea*), phacelias (*Phacelia* spp.), coreopsis (*Coreopsis bigelovii*), desert monkeyflower (*Mimulus bigelovii*), purple mats (*Nama demissa*), desert star (*Monoptilon bellioides*), suncups (*Camissonia* spp.), desert-chicory (*Rafinesquia neomexicana*), pincushion flowers (*Chaenactis* spp.), lupines (*Lupinus* spp.), gilias (*Gilia* and *Linanthus* spp.), evening primroses (*Oenothera* spp.), and many, many more.

On higher slopes, creosote bush scrub intermingles with the weird and wonderful Joshua tree (*Yucca brevifolia*) and a variety of shrubs and cacti. Higher yet, Joshua trees mix with and are replaced by pinyon pines, junipers, and sagebrush scrub. Only on the very highest desert mountains, such as along the western edge of Death Valley or in the New York Mountains, are there forests of taller conifers. Especially striking there are the open forests of bristlecone pine (*Pinus longaeva*), reputedly the world's longest-lived trees.

To the south of the Mojave Desert and across the Transverse Ranges lies California's winter-warm Sonoran Desert, an extension of the vast desert of southern Arizona and northwestern Mexico. This desert experiences relatively mild winters because it lies close to or below sea level. In addition to creosote bush scrub, large dry desert

Desert oasis with pond.

washes are lined with a woodland of smoke tree (*Psorothamnus spinosus*), desert-willow (*Chilopsis linearis*), desert-ironwood (*Olneya tesota*), catclaw acacia (*Acacia greggii*), mesquite (*Prosopis glandulosa*), and palo verde (*Cercidium floridum*). (This community is poorly developed in the Mojave.) Desert oases occur where fractured fault lines allow water to seep to the surface year round. These oases are home to California's beautiful fan palm (*Washingtonia filifera*) and an array of other riparian trees, shrubs, and flowers.

Joshua trees are conspicuous by their absence in the Sonoran Desert. Instead there is the odd, slender, cactuslike, whip-stemmed ocotillo (*Fouquieria splendens*), growing

with a large assortment of cacti and other succulents, including Mojave yucca (*Yucca schidigera*), beavertail cactus (*Opuntia basilaris*), barrel cactus (*Ferocactus cylindraceus*), chollas (*Opuntia* spp.), hedgehog cactus (*Echinocereus engelmannii*), and desert agave (*Agave deserti*).

This overview of California plant communities is highly condensed, and intricate detail has been omitted, but this is the essence of where the plant communities occur and how they're composed. Hopefully, it's clear that there is an immensely varied palette of trees, shrubs, annuals, perennials, bulbs, ferns, and ground covers from which to choose in creating beautiful gardens appropriate to any site. We have selected twelve of these communities to portray the parts of California we feel are best suited to creating beautiful gardens.

Floriferous ocotillo (*Fouquieria splendens*).

Dramatic rosettes of chaparral yucca (*Hesperoyucca whipplei*).

PLANT COMMUNITY–BASED GARDEN DESIGN
A Garden-Making Ethic for California
by Alrie Middlebrook

Why do I speak of a garden-making "ethic"? Gardens are for planting favorite roses, spending time with the family, therapeutically pulling weeds, and mowing the lawn. Ethics, on the other hand, are about riding a bicycle to work or running a 10K race that benefits leukemia research—aren't they? What do gardens have to do with ethics?

Plenty. When I design a garden, I follow three primary rules, which, taken together, I call my garden-making ethic. The first is to create a place of beauty. I do this by applying design principles like texture, balance, color, repetition, and scale. Frequently, I will incorporate pieces of found or manufactured art, both items collected over the years and new finds that fit the space well. Because art has the power to engage and entertain, I think of the garden itself as a piece of art, an ever-changing work in progress.

The second rule is to create a space with meaning. All of us bring both personal and cultural history with us to our gardens. Childhood memories—picking fruit with our grandfather, planting roses with our mother, even pulling weeds and making compost heaps—play an important part in garden creation. Cultural history does as well. We may have grown up with an Asian-style garden or one that featured cacti and succulents. We may respond to English borders. My gardens need to resonate with the owners' backgrounds, so they feel comfortable and provide a sense of being-at-home.

The third rule is to study the surrounding natural community, imagining what it was like before it was disturbed. Was the local landscape oak woodland or mixed-evergreen forest, or perhaps a riparian woodland or grassland meadow? I learn not only the plants and

LEFT: Blue foothill penstemon (*Penstemon heterophyllus*) in chaparral garden. Photograph by S. Ingram. ● ABOVE: Mixed-evergreen woodland with a gravel path.

flowers of the area but also the fauna, including the birds, insects, and butterflies. I observe the angle of the sun, the slope of the property; I examine the soil and note how water moves across the site. In my designs, I try to incorporate all the natural advantages of the space so that it is not only beautiful, but also ecologically sustainable. I also try to reuse, renew, and recycle elements in the garden-making process.

Marrying cultural and ecological appropriateness can be a challenge, but that is part of the pleasure of garden design. If I use exotic plants, for example, I try to display them in containers or raised beds, setting them apart from the native flora of the site. By grouping them together, especially geographically, the designer can emphasize their difference and uniqueness to good effect. This technique works not only visually, but also horticulturally, by keeping natural systems intact.

Good garden design expresses a harmonious approach to living on the Earth, striking a balance between natural ecology, cultural variation, and personal aesthetics.

Creating a garden that conforms to this ethic involves six main steps: evaluating the physical site; selecting the plant community; designing the garden; creating the hardscape; building the garden; and finally, maintaining it. I cover each of these steps in detail in what follows.

Phase One: Evaluating the Physical Site

When surveying the garden site as well as the local surroundings, you need to ask some basic questions. For instance:

- ◆ What are the existing topographical features (e.g., slopes, trenches, hillocks)?
- ◆ How is the site oriented?
- ◆ How does the property drain, and where does it drain to?
- ◆ Is the site near a natural feature such as a creek, mountain, or forest?
- ◆ Are there any still-intact natural plant communities nearby?
- ◆ Are there existing large plants and trees that will be retained?
- ◆ What types of landscape materials already exist at the site? Can they be re-used in a new design?
- ◆ What kinds of garden landscapes do adjacent or neighboring properties feature? Will these influence the design of the site?

Keep the answers to these questions in mind as you move on to the next step: deciding on plants.

Phase Two: Selecting the Plant Community

- ◆ First, determine the natural plant community of the site prior to urbanization,

Native hillside garden under redwood trees (*Sequoia sempervirens*). Photograph by S. Ingram.

for instance, oak woodland, grassland, or chaparral. If these plant communities are compatible with the site, assess their water, slope, and light requirements; this will help you decide how to group new plantings. Consider whether additional landscaping features could be included to promote the plants' survival. For example, for a seep or wetland planting, a water feature could be added; and a dry creek bed (which becomes a flowing creek during winter rains) would be a good addition if a riparian community is being planned. Try to re-create the physical qualities of the plants' natural environment. Chaparral species, for instance, evolved on rocky slopes with hot southern and western exposure. You can replicate this situation on a smaller, more manageable scale in the home garden by building small berms, twelve to eighteen inches high, and adding rocks and boulders.

- If existing trees will be incorporated into the garden design, select and create an understory consistent with the trees and their natural community. This applies whether the trees are native or nonnative. For example, if you will be working with riparian species such as sycamore, maple, and alder, choose a riparian planting palette for the understory, which will have similar water needs and tolerate the shade provided by the deciduous trees overhead.
- A single garden can sustain several plant communities, depending on the microclimates of the site. The size of the planting schemes can be small or large; this, too, is determined by existing microclimates. It is best to keep the plants from each community together. However, if you choose to mix and match, select plants that have compatible water and drainage requirements—for example, those of chaparral and oak woodland, or those of riparian and redwood forests.

Phase Three: Designing the Garden

Once the plant communities have been selected, various factors will influence the final garden design.

- *Budget.* The process of discovering a realistic garden-making budget can be complicated. I usually begin by designing a master plan that includes many of the dreams and wishes of the owner. When a real price tag is attached to each item, the owner is then able to choose those features that are most important. Often, a project is carried out in phases as resources become available.
- *Owner's cultural heritage.* I try to incorporate specific references to the owner's culture, whether it was growing up on a farm in the Midwest or in a high-rise in Hong Kong.

- *Intended use of the garden*. Whether the garden will be used for recreation, quiet contemplation, entertaining, cut flowers, sustainable edibles, wildlife habitat, a children's play area, a display or collector's garden, or for a combination of these purposes, will influence the ultimate design.
- *Time line of maturity*. Gardens can mature very quickly if fast-growing plants are selected, while others may take years to reach full maturity. Whether people want a design that will look stunning in five years when they plan to sell their home or, facing retirement, wish to have a low-maintenance garden that will last twenty years, plant choices will vary based on owners' needs.
- *Architectural style of house*. This factor will determine which construction materials, hardscape features, and garden amenities are most appropriate. For

TOP: Purple needlegrass (*Nassella pulchra*) and Idaho fescue (*Festuca idahoensis*) bordering a lawn of *Carex pansa*. Photograph by S. Ingram. • BOTTOM: Creeping red fescue (*Festuca rubra*) living roof with custom iron fence detail.

example, Arts and Crafts-style homes feature extensive use of natural rock, as well as arbors and porches that create outdoor rooms. These elements provide opportunities to design other garden structures and focal points that are compatible. Other architectural styles that can strongly influence garden design are Victorian/Cottage, Spanish/Mediterranean, Craftsman/Woodland, Modern/Minimalist, and Traditional/Formal. A strong cultural tradition supports certain combinations of home and garden in these cases. People tend to be comfortable with a look they have seen before and for which they can easily visualize the textures, forms, and colors that go into its creation.

Typically, our home represents our life's investment. Most people who consider changing their garden's style from the traditional lawn, foundation shrubs, and colorful annuals to a natural ecological garden want to be assured that their new garden will be visually appropriate for the neighborhood, increase the overall value of their property, and prove a wise investment over time. The single most compelling force that enables people to switch is the enchanting beauty of our native flora. The opportunity to express beauty with style and panache makes change a welcome alternative.

◆ *Inclusion of art.* If pieces of art will be incorporated into the garden design, will it be collected by the owner? Created by the owner? Created by commissioned professionals? How does the art relate to the architectural style of the house, or to the overall theme of the garden?

Although many people think the inclusion of art in a garden will be too expensive, or potentially controversial, or an invitation to thieves, understanding what art is may put these fears to rest. Art is an expression of personal creativity. That is all that needs to be considered.

One medium that lends itself to garden art and is easy for amateurs to create is mosaic. Entire families can come together to create mosaic tables, fountains, decorative containers, and murals, possibly using motifs of plants and animals that are found locally. For oak woodland gardens, for example, acorn finials, acorn fountain elements, and oak tree ornamental gates make lovely accents. If clients prefer to purchase art for their gardens, many artists show their work at open studios, where it can be acquired at reasonable prices. Old found objects can also be used in artful ways.

◆ *Use of the principle of ecological succession.* When disturbances occur in na-

Idaho fescue (*Festuca idahoensis* 'Siskyou Blue') with accent boulders.

ture, such as fire, mudslides, or earthquakes, new plants move into the devastated landscape. These early species, which may include grasses, short-lived perennials, and annuals, are fast growing. They help prepare the soil for the longer-living shrubs and trees that will eventually replace them. When I create a new garden, I include both of these steps simultaneously.

The process is easy. Start with slower-growing woody shrubs, trees, and ground covers, selecting compact species and noninvasive plants that will not create excessive garden waste or require repetitive, labor-intensive care regimes. Space these plants at densities that will allow them to grow to their mature size without overcrowding or overwhelming other species. Then fill in between these new plantings with shorter-lived species, like bunchgrasses, annual wildflowers, and fast-growing shrubs and perennials. Using the principle of succession you can create a garden that

TOP: Coastal sage garden with manzanita (*Arctostaphylos* sp.). • BOTTOM: California buckwheat (*Eriogonum fasciculatum*) showing fall color.

looks quite mature within a few months after planting, but that also has the strong bones to remain a beautiful garden for twenty years.

- *Appropriate densities.* Many professionally designed gardens are overplanted. This is especially true of commercial properties, such as shopping centers and hotels, as well as upscale housing projects. Although the developers want to create a sense of immediate lushness and opulence, what they have in fact created is a maintenance nightmare. All too quickly, the plants will need constant pruning and hedging, and they will require large amounts of water and create more garden waste. A look of lushness can be achieved at far less expense simply by planting at appropriate densities, leading to far lower maintenance costs in the future.
- *Turf issues.* The lawn culture is a relic of our agrarian past, associated with landowners who chose to devote a portion of their property to prestige rather than food production. In other words, they were wealthy enough to forgo animals grazing in their front yards. In California, where new housing developments are being built in more arid regions of the state, the ubiquitous lawn can account for up to 75 percent of a typical household's water use. Not only that, but pumping water throughout the state consumes large amounts of energy. This solution for landscaping is untenable in a state whose population is expected to be fifty million by the year 2020. Where will all the water for all these lawns in the desert come from?

Fortunately, there are many attractive alternatives to turf that require far less water, such as native bunchgrasses and native ground covers. Another option is replicated grass; made of recycled plastics, recycled tennis shoes, and sand, it creates a grasslike surface that is guaranteed for ten years. The first time I walked on sports turf, I thought it was real grass. You can also replace a water-guzzling lawn with a natural wildflower meadow, using bunchgrasses, annuals, bulbs, and perennials.

Phase Four: Creating the Hardscape

- Channel rainwater and runoff into the built landscape—but avoid directing it into the street or storm drain. Use permeable surfaces whenever possible, such as clay pavers, and flagstone on decomposed granite, and pervious concrete. (After the burning of fossil fuels, the manufacture of cement is the number-two contributor to global warming in industrialized nations.)
- Create catchments to hold precious rainwater, as well as swales and rivulets to divert and slow seasonal runoff. In gardens I have designed, dry creeks become seasonal creeks during the winter months. Downspout water is diverted

with a buried flexible tube that empties into the dry creek. Riparian species planted in and near the creek happily respond to the seasonal flow of water.

- ◆ Recycle existing concrete and outdated materials; rather than transporting demolished materials to landfills, reuse them on-site. If you choose to retain concrete, resurface it.

- ◆ Build new hardscape features from materials that use minimal processing or contain a high content of recycled material, such as Trex or Weatherbest wood composites, created with recycled plastic and wood pulp. Rammed earth, PISÉ (pneumatically impacted stabilized earth), poured earth, straw bales, recycled tires, salvaged objects, and other manufacturers' recycled wood, glass, and plastic products are sustainable alternatives for traditional construction materials.

- ◆ Use local or indigenous materials whenever possible, in part to cut down on excessive transportation and delivery costs.

- ◆ Employ eco-engineering principles, such as using plants to stabilize a steep bank, water retention basins that double as design features, and drainage diversion options that retain rainfall for irrigation use. Plant terraces can substitute for costly retaining walls, and plantings employing the principle of succession will help control erosion.

- ◆ Select local artists and craftsmen for one-of-a-kind garden amenities. Building collaborations with local creative people can add artistic synergy to the garden-making process.

Living roof of creeping red fescue (*Festuca rubra*) at Esalen Institute.

Masses of island alumroot (*Heuchera maxima*) in an oak woodland garden.

◆ If possible, include an active water feature in the garden. Birds love water. Water noise is soothing and masks engine sounds. The movement, sound, and fluid energy of water will attract people to the garden. Water is more sensual if it trickles, less effective if it gushes.

Phase Five: Building the Garden

◆ Disturb soils as little as possible during the construction process. If you import soils, use soil that has not been contaminated or exposed to weed seeds. Ideal imported soils come from urban excavated areas, such as a hole dug for a swimming pool, house foundation, or basement. Whenever possible, create berms utilizing site soil. If incorporating a dry creek bed as a design feature, for example, be sure to retain the soil excavated for the watercourse and use it to build berms.

Avoid compaction and cut-and-fill practices. Work with existing grades as much as possible. If you must disturb soils, use hand tools instead of machines. Never rototill; the less soils are disturbed, the fewer weeds will germinate.

◆ If you choose to replace a large expanse of turf with a bunchgrass and wildflower meadow, hydroseeding is especially effective. To reduce weeds, before planting, cut any existing weeds and remove. Then spread sheets of thick plastic on the meadow area. In summer, temperatures should reach 140 degrees under the plastic. After two months, remove the plastic and hydroseed immediately. If the area is small (as on a typical suburban lot), hand seeding is more practical, but more weeds will germinate than with hydroseeding. In either case, hand weeding is necessary for the first two to three seasons, though this task diminishes greatly after the first year, especially if you weed diligently and continue to reseed in areas that were slow to germinate. By the third season, the native grasses will have outcompeted the weeds.

◆ Retrofit existing irrigation to drip, limiting overhead spray to meadow areas. Install drip lines underground or buried beneath a thick layer of organic mulch. Never install overhead spray in narrow planting strips. It is crucial that nursery-grown native plants receive a conventional water regime for the first two seasons.

◆ Don't amend soils for native plants. (Exceptions are hand-seeded meadows. In this case, it is helpful to add compost to the seed bed.) If soil is heavy clay, create mounds for proper drainage. Compact mounds before planting.

◆ When planting a native garden, assuming there is adequate irrigation, expect a plant loss rate of up to 20 percent. This rate is determined by many factors:

the quality of the plant, the general health of the root system, the prevailing weather conditions when the plant was installed, the distance from the garden site to the nursery where the plants were grown, and the species of the plant. If the temperature is 90 degrees or above during the first few weeks after the plant is installed, the loss rate will be greater. Do not plant if temperatures are over 90 degrees or if the soils are saturated from continued rains. To reduce plant loss, apply a thick layer of mulch immediately after planting.

◆ Practice fuel economy at all stages of the construction and maintenance process. Select electric-powered equipment over gasoline power. Best choice is hand-powered equipment.

◆ Apply mulch generously (three to four inches). Use mulches and composts made from recycled curbside waste. We use ProChip, a product manufactured by BFI made from recycled wooden pallets, discarded used wood, and other curbside organic waste. If you use mulch supplied by tree services, be sure that it is free from invasive weeds and seeds. Avoid mulch from eucalyptus, bay, and walnut trees. These species are allelopathic; they produce chemicals that inhibit the germination and growth of other species under their canopy.

◆ For gopher control, cover the entire planting area with a grid of chicken wire

Giant wild rye (*Leymus condensatus* 'Canyon Prince') in a coastal planting at Rancho Santa Ana Botanic Garden.

prior to planting, securing wire along the perimeter; use wire clippers to cut holes for individual plants. Alternatively, install individual chicken-wire baskets for each plant. Cover with organic mulch, or for a meadow planting, top dress with a recycled organic compost.

◆ For deer control, construct bird netting cages, using a structure of bamboo stakes or other lightweight materials. Build the cages to the approximate size the plant will be at two years, at which time the protective cages can be removed. By then, the plant will be established and deer can browse it without damage, or the plants—whose taste changes as they grow—may well have lost their appeal altogether. Usually, deer will not browse salvias, irises, or ferns. At present, California has an inflated deer population due to the availability of food and water from excessive irrigation, overplanting of exotics, and loss of predators. Creating more native gardens will help restore balance to the deer population. Deer evolved with the native plant communities and occupy an important niche within that ecosystem.

Phase Six: Maintaining and Managing the Native Garden

◆ Whereas exotic gardens require repetitive maintenance practices such as lawn mowing, constant pruning and hedging of fast-growing trees and shrubs, and the frequent purchase and planting of short-lived perennials and annuals— leading in the end to enormous garden waste—native gardens are far less demanding. Each one will have a different management protocol depending on the site, the owner's tastes, and the plant communities chosen. For example, if you replace a conventional lawn with native Californian bunchgrasses, you may choose to cut the meadow twice a year or four times a year, depending on whether you prefer a natural look or a more formal appearance.

◆ Plan for the garden to become self-sufficient and free from irrigated water after two years, except during extreme droughts. This can be done gradually, by reducing irrigation after the first few months, and then more during the second year. Exceptions are riparian, montane meadow, wetland, and redwood forest communities, which require some summer water even after they have become established. If you prefer a bunchgrass meadow to remain green during summer, it will need additional water. With multiple plant communities in one garden, it is important that each community's water regime be monitored during the two-year establishment period. Weather, microclimates, and growth habits affect water use.

◆ Management of native meadows is the most labor-intensive part of native gardening. Weed control—a critical task for the first two seasons—is best

achieved through hand pulling. Knowing how to recognize nonnative weeds and grasses before they flower is therefore imperative. By the third season, if sufficient time and effort was devoted to weed control, the natives will have successfully outcompeted the exotics. For the first two seasons, too, cut grasses after they have flowered, and continue to reseed sparse areas.

◆ For small urban spaces, it is important to manage the anticipated growth of trees and shrubs in the first few years of the garden's growth. It may be necessary to encourage verticality of tall shrubs in narrow spaces, for instance, or to espalier a plant to a south-facing wall.

◆ If you want to rejuvenate plantings of perennials, simply remove the spent ones and add new plants, taking care to replace the drip lines and to mulch. Flowering perennials can be cut back for a second bloom or allowed to set seed for wildlife. With good management practices, perennials may live three to five years. Many will reseed or naturalize.

◆ After three to five years, certain chaparral species may grow woody and sparse, or become out of scale with the surrounding vegetation. One solution is to severely reduce their size. Most species (not some manzanitas, however), if cut to the ground, will sprout from the base. This technique will rejuvenate the plant and extend its life many years. If you have several plants

TOP: Chaparral garden with a hillside arbor. Photograph by S. Morris. • BOTTOM: Terraced grassland garden in a Mediterranean setting.

Idaho fescue (*Festuca idahoensis* 'Siskiyou Blue') in a bluff garden. Photograph by S. Holt.

of a single species of chaparral in a mass planting, you may want to selectively prune them over a couple of seasons, until the whole area has regrown. Prune during the fall, but before winter rains begin.

♦ Treat pests with detergent or Safer's Soap instead of using pesticides and insecticides, or simply cut off infested and diseased parts of plants and dispose. If necessary, remove the entire host plant. The best way to avoid such problems is not to buy plants that are susceptible to pests and diseases. Since most native plants have evolved in competitive, complex communities, their systems resist indigenous insects. Insects that have been introduced from other exotic ecosystems or exotic plants that attract harmful pests can, however, bring infestations into the garden.

Organization of the Book

Of the numerous native plant communities and their variations, we've chosen twelve that best represent habitats and climates in which most Californians garden. In a book this size, it is impossible to be all-inclusive, so the plant selections for each community have been driven by considerations of aesthetic design and availability. Certain species, though perhaps difficult to find in the trade, were chosen for their overall appropriateness and beauty; many other species are easily found in a variety of nurseries. To learn about sources of natives, turn to Appendix 1, where we've listed the majority of nurseries that deal in California natives.

Because many species live in several different communities, some of our selections appear more than once—especially in the lists of additional plants to try, given at the end of the species descriptions for each plant community. Where a species is described in full, its inclusion is driven by design considerations as well as appropriateness to a particular habitat type.

Each chapter is organized in the following fashion:

♦ Introductory paragraphs describe the natural plant community, as well as its climate and setting.

♦ We provide a design for a garden based on the plant community in question. Each design includes four sections: design notes; scope of work required; appropriate plants; and a planting plan.

Dudleya ssp. in a container.

- ◆ We then briefly describe 20 to 35 different species that are well suited to each plant community. Cultural requirements and methods of propagation are specified. The species lists are organized according to horticultural categories such as *trees, shrubs, grasses, perennials, ground covers, ferns, bulbs,* and *annuals*.
- ◆ We provide guidelines for creating a community-based garden, and comment on basic design elements and considerations.
- ◆ Additional species to try are listed.
- ◆ We recommend places to go to visit the plant community in its natural setting. We've tried to select a variety of locales throughout the state; however, some plant communities are best developed in limited areas—the redwood forests of far northern California, for example, or the coastal sage scrub of southern and central parts of the state.

Chaparral planting and grassland elements.

CHAPTER 1

BLUFFS AND CLIFFS: ELEMENTS FOR A ROCK GARDEN

Though seemingly inhospitable, bluffs, cliffs, and rock faces are home to a host of intriguing and attractive plants. Here you see draped, matted, or tufted plants clinging to precipitous coastal bluffs or tucked into crevices of cliffs in the foothills. Each species is fitted to its environment in its own special way, sporting adaptations that allow survival in the face of frequent winds, rapid runoff of water, and minimal soil. Such plants are excellent candidates for domestic rock gardens. (The exceptions are plants from the alpine environment above timberline, which, though intriguing and beautiful, are unlikely to succeed in low-elevation gardens because of their need for long dormancies and winter cold.)

Plants that grow on coastal bluffs experience some compensations to their difficult environment: temperatures are mild year round, with few winter frosts and few hot summer days, resulting in a long growing season. Although summers are usually rainless, coastal bluffs are often bathed in dense fogs that dramatically reduce transpiration. By contrast, foothill cliffs get no relief from hot, dry summers, with a rainless period that lasts from late April through late fall. Plants here are well adapted to drought and prefer full sun year round. Their growing season coincides mostly with the winter and spring rains, although a few tough plants continue to bloom as late as fall.

Most bluff plants adapt by having extensive roots to anchor them in place and store water, by growing as dense mats or cushions to avoid the brunt of winds and to retain moisture, and by possessing fleshy or hairy leaves that hold extra water and reduce water loss. Each species has its own look, and combinations of these plants—both in blossom and out (many flower from late winter through early summer)—create beautiful tapestries in carefully designed rock gardens.

In addition to a wide variety of colorful perennials and bulbs, coastal bluffs are also home to sprawling woody ground covers, some of which grow into taller shrubs inland, away from winds and under the protection of forests. In a rock garden, these woody plants can create a pleasing backdrop for matted herbaceous perennials.

LEFT: Rugged coastal cliffs at Pt. Lobos State Reserve in Monterey County. • ABOVE: Coast dudleya (*Dudleya farinosa*).

Inland cliffs feature an array of small bulbs, bunchgrasses, cushion-forming perennials, and woody ground covers. They are also home to a range of attractive small ferns, many of which go dormant in summer drought.

Creating a Garden

When you're ready to create your own design, try a pleasing mix of several different perennials selected for their bloom time, color, and leaf design and complemented by one or two kinds of grasses, succulents, and ferns. Bulbs can be tucked into corners between other plants as an element of surprise. Start with a few woody plants to create the foundation and backdrop, then add matted and cushion-forming wildflowers accompanied by clumps of succulents for contrast, native bunchgrasses for texture, and pock-

TOP: Bluff garden near the coast. Photograph by A. Owens. • BOTTOM: California sagebrush (*Artemisia californica*) with accent boulders.

ets of spring-flowering bulbs for color. Inland, you'll also want to include some of the beautiful rock ferns.

Select from the several species described below. Be sure to keep in mind the origin of the material: coastal bluff plants prefer cool, foggy summers; plants from foothill cliffs do best with full summer sun. If you happen to live in a place between these extremes, try plants from both habitats. For strictly coastal garden sites or areas with blistering summer heat, it's better not to mix material from the two different habitats. Each entry indicates whether the plant is adapted to coastal or inland conditions.

TOP: Coastal bluff garden with sagebrush, coyote brush, pines and cypress. • BOTTOM: Bluff garden with riparian feature.

Existing vegetation coastal bluff and coyote brush on disturbed hillsides

Green roof planted with grasses, perennials and annuals (seed)

Upper patio on roof constructed with stained concrete planks

Glass block sky windows

Mosaic tile on concrete

PISÉ or poured-in-place concrete house

Garage door, aluminum framed with Trex panels

Flagstone path to roof patio

"Gravelpave" system installed California Gold gravel

Design Notes

Coastal bluffs, with high winds and frequent winter storms, lend themselves to earth-sheltered homes. The garden and house shown in this drawing will incorporate plants from the coastal bluff community.

In order to build the house on this site, significant excavation must occur. Construction of the home can utilize site soil mixed with cement, liquefied and poured in place to form rigid, durable walls. All soils should be slowly removed, taking care to maintain the relationships of soil types, layers, distributions, and microorganisms, to the extent possible. The top layer, with plants intact, should be removed to a site nearby, preferably with similar directional exposure. Upon completion, soils should be replaced in the same order in which they were extracted. In certain areas, it is not possible to preserve existing flora; the design therefore includes a native bluff garden on south- or west-facing slopes, with species such as California sagebrush and glorymat, especially near the outdoor patio, where it can show off its lovely blue flowers. Before construction begins, cuttings of existing natives from the site can be propagated in a small greenhouse; the plants that grow from these cuttings will then be planted in the new garden.

Rendering of featured garden.

This earth-sheltered house will also incorporate living roofs, a sandwich of concrete planks, rigid insulation, a waterproof membrane, and a drainage layer of sand or gravel and soil. On top of the kitchen there will be a rooftop patio of colored concrete planks, shielded from the wind by three coastal bluff meadows on the barrel-vaulted, living roofs. The roof garden will be accessed by a curved flagstone stairway. As one climbs the steps,

a rock garden featuring existing boulders and coastal bluff plantings unfolds, with plants draped or matted and shaped by the wind.

The bones of the rooftop garden will be provided by massed woody shrubs of readily available species, such as coffeeberry, kinnikinnick, and coyote brush. Flowering perennials include golden aster; coast buckwheat; sulfur buckwheat, an inland species that may not like the fog but has striking yellow

flowers; seaside daisy; coyote mint, another inland species; and bluff lettuce, which naturally occurs in the crevices of rocks. The barrel-vaulted roofs will be planted with Cape Mendocino reed grass, but it is possible to use tufted hair grass or Nutka reed grass—or a combination of the three, although the larger Nutka reed grass will dominate small spaces. Tucked in between a few lightweight lava rocks, seathrift will provide a nice accent. Roof gardens such as these can be seeded with annuals as well.

An original mosaic made of broken glass and ceramic tiles will decorate the entryway, featuring bluff lettuce as a design motif because of its red-tipped leaves. Nearby, a hand-thrown basin decorated with a coffeeberry mosaic will serve as a birdbath, coffeeberry being a favorite cover plant for the white-crowned sparrow that makes its home in the coastal bluff plant community.

TOP: Seaside daisy (*Erigeron glaucus*) with bearberry (*Arctostaphylos uva-ursi*) and iris. Middlebrook Gardens demonstration garden. • BOTTOM: A natural bluff garden at Pt. Lobos includes seaside daisy and coast buckwheat (*Eriogonum latifolium*).

Scope of Work

- Excavate the site soil for building construction. At completion of house, replace soil and plants.
- Build a temporary greenhouse and propagating facility on the site.
- Create kitchen-top roof garden using sand or gravel and soil.
- Install grass seed and plugs in roof garden; sow wildflower seed in roof garden meadows.
- Install driveway entry with compacted base rock, Gravelpave* system, and California Gold gravel.
- Install flagstone stairway. Utilize existing boulders to create small retaining walls for stairway and rock garden.
- Install drainage system adjacent to stairway.
- Install irrigation and plantings.
- Install ProChip** mulch on planted areas.
- Create entry column and install glass or ceramic mosaic.

* Gravelpave is a manufactured rigid plastic honeycomb grid system that holds gravel in place and prevents ruts and low areas from forming. It is manufactured by Invisible Structures, Inc. (www.invisiblestructures.com).

** ProChip is a mulch made from recycled wooden pallets and other organic curbside waste. It is available at BFI Industries.

Plant List

Symbol	Botanical Name	Common Name
A	*Artemisia californica* 'Canyon Gray'	California sagebrush
B	*Armeria maritima*	seathrift
C	*Calamagrostis foliosa*	Cape Mendocino reed grass
D	*Ceanothus gloriosus*	glorymat
E	*Calamagrostis nutkaensis*	Nutka reed grass
F	*Deschampsia caespitosa* ssp. *holciformis*	tufted hair grass
G	*Baccharis pilularis* 'Pigeon Point'	dwarf coyote brush
H	*Rhamnus californica* 'Seaview'	dwarf coffeeberry

Annual seed mix for roof garden meadow to include:

Castilleja exserta	owl's clover
Eschscholzia californica maritima	California coastal poppy
Lupinus nanus	sky lupine
Nemophila menziesii	baby-blue-eyes

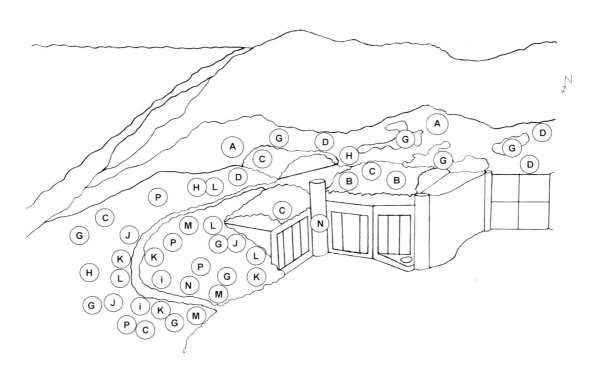

Coastal Bluff Rock Garden

i	*Heterotheca sessiliflora*	golden aster
J	*Eriogonum latifolium*	coast buckwheat
K	*E. umbellatum*	sulfur buckwheat
L	*Monardella purpurea*	serpentine coyote mint
M	*Erigeron glaucus*	seaside daisy
N	*Dudleya farinosa*	bluff lettuce
G	*Baccharis pilularis* 'Pigeon Point'	dwarf coyote brush
P	*Arctostaphylos uva-ursi* 'Pt. Reyes'	kinnikinnick
H	*Rhamnus californica* 'Seaview'	dwarf coffeeberry

Plants to Use

Background Woody Plants

Dwarf coyote brush (*Baccharis pilularis*), COASTAL OR INLAND.
Mounded woody ground cover to around three feet high with an equal or greater spread; fast growing. Needs periodic shearing to keep it dense and low. Closely set, glossy

Featured garden key.

green, rounded, fragrant leaves. Small pale yellow flower heads in fall. (Plant males only; female plants produce hairy white seed clusters that create a mess.) Minimal water. Propagate by layering or cuttings. Maintain by frequent tip pruning to keep the low profile and improve density. Plants may need to be restarted every few years as centers often die out over time. The taller version of coyote brush (var. *consanguinea*) can also be shaped into a sculptural element.

Glorymat and Hearst ceanothus (*Ceanothus gloriosus* var. *gloriosus* and *C. hearstiorum*), COASTAL. [Inland substitute: little-leaf ceanothus, *C. foliosus*]

A pair of evergreen, woody ground covers from three to eight inches high, rooting as they grow. Glorymat has glossy, hollylike leaves and rounded clusters of blue flowers; Hearst ceanothus has tiny, narrow, sticky, wrinkled leaves and similar clusters of dark blue flowers. Both bloom in midspring. Propagate by layering or cuttings. Shear periodically to keep trim. Center part may die out and plants sometimes lose vigor with age; replace them when those conditions persist.

Garden Design Note: Faster growing than manzanita ground covers.

California coffeeberry (*Rhamnus californica*, dwarf cultivars such as 'Seaview'), COASTAL OR INLAND.

Sprawling woody shrub or ground cover to four feet tall; needs periodic clipping to maintain its low form. Dark evergreen, narrowly elliptical to oblong, finely toothed leaves on red new shoots. Insignificant yellow-green, starlike flowers in late spring to early summer, followed by dark purple fruits attractive to wildlife. Propagate from hardwood cuttings. As with other dwarf forms, shearing once a year to maintain dense growth and prevent legginess is a good idea.

Garden Design Note: Performs well in sun or shade. Layer California coffeeberry with 'Eve Case', 'Mound San Bruno', and 'Seaview'.

Kinnikinnick and Hearst manzanita (*Arctostaphylos uva-ursi* and *A. hookeri* var. *hearstiorum*), COASTAL. [Inland substitute: Rincon manzanita, *A. stanfordiana* var. *repens* and Little Sur manzanita, *A. edmundsii*]

Near-prostrate, woody, evergreen ground covers that drape over banks and root as they grow. Dull, dark green, rounded leaves; red bark; and tiny white bell-like flowers

TOP: Leaves and flowers of Hearst ceanothus (*Ceanothus hearstiorum*).• BOTTOM: Branches of coffeeberry (*Rhamnus californica*) with flowers.

in midspring, followed by vivid red berries. Several cultivars are available for kinnikinnick. Hearst manzanita grows even lower, with tight, overlapping, bright green leaves. It blooms a bit later and is slower growing but is the tidiest of the low manzanitas. Propagate from rooted stems or cuttings. Plants need little maintenance after they have filled in; be sure to allow plenty of room for them to spread.

Garden Design Note: Tidy, lush green ground cover suitable for a formal garden.

California sagebrush (*Artemisia californica* 'Canyon Gray'), COASTAL.
Semiprostrate woody ground cover that arches gracefully over banks. Sage-scented, grayish, finely divided foliage (may defoliate in hot dry summers); slender spikes of tiny, pale yellow flower heads in fall. Sensitive to cold, wet winters. Propagate from layered branches. Shear off flowering stalks when flowers have finished. Centers may die out in time.

Garden Design Note: Excellent massed on a hillside with waves of sticky monkeyflower and coyote brush interspersed.

Living roof overlooking a coastal bluff community.

Cushion-forming and Matted Perennials

Dune sagewort (*Artemisia pycnocephala*), COASTAL. [Inland substitute: California sagebrush, *A. californica*]

Tufted clumps of grayish, wool-covered, pinnately divided leaves forming mounds to eight inches high. Leaves are unscented and usually evergreen. Spikes of insignificant, yellowish green flower heads in summer may be trimmed off. Handsome for texture and color of foliage. Propagate from semihardwood cuttings. Shear off flowering stalks if their appearance detracts from the overall effect; also remove old flowering stalks after flowers have faded.

Garden Design Note: Excellent in a woven mosaic of 'David's Choice' and dwarf coyote brush (*Baccharis pilularis* 'Pigeon Point').

Golden aster (*Heterotheca sessiliflora*), COASTAL OR INLAND.
Tufted clumps of elliptical, shaggy-haired, dull green to gray green leaves on mounds six to eight inches high. Long succession of bright yellow, daisylike flowers in summer and early fall. Propagate by division or seed. Needs little maintenance, though clumps can be periodically divided to thin the plants when growth is too dense.
Garden Design Note: Effective in masses, interspersed with seaside daisy.

Woolly aster (*Lessingia filaginifolia* var. *californica*, COASTAL, and var. *filaginifolia*, INLAND).

The coastal form, var. *californica*, is a semiwoody, low, evergreen ground cover a few inches high with elliptical, silvery, wool-cov-

TOP: Dune sagewort (*Artemisia pycnocephala*). • BOTTOM LEFT: Bluff garden with boulders, lupine, and dune sagewort. • BOTTOM RIGHT: Coast dudleya (*Dudleya farinosa*).

ered leaves. A particularly attractive cultivar of this variety is 'Silver Carpet'. The inland form, var. *filaginifolia*, is a woody perennial with similar leaves but grows to two feet high. Both produce asterlike flowers with red-purple rays and yellow discs in summer and fall. Propagate by rooted sections, divisions, or seed. Little maintenance needed. Garden Design Note: Appropriate for a cottage garden, as the stems branch and are covered with small pink asters that ramble over the ground.

Seathrift (*Armeria maritima*), COASTAL.

Rosettes of narrow, grasslike leaves attached to a stout taproot. Naked stems from

four to 10 inches tall carry pompons of pink or rose-purple, long-lasting blossoms from late spring to early summer. May increase by branching to produce several rosettes. Propagate from offsets or seed. If plants become leggy after a few years, cut off clumps to propagate, and cut back stems so the new plants will look trimmer.

Garden Design Note: Makes a great border plant, nestled among rocks.

Coast buckwheat (*Eriogonum latifolium*),
COASTAL.

Low, sprawling cushions of spoon-shaped, gray-felted leaves on plants to a foot high. Large heads of white to deep pink blossoms in summer on naked stalks several inches above foliage. Flowers attractive to many different pollinators. Propagate from seed or

cuttings. Remove old flowering stalks after seed has set.

Garden Design Note: Mass with boulders. Flowers display many colors over an extended bloom cycle.

Sulfur buckwheat (*Eriogonum umbellatum*),
INLAND.

Mounded woody cushions or small shrubs from six inches to two feet high according to form. Dense, spoon-shaped, dull green to grayish basal leaves. Broad, umbel-like masses of sulfur yellow flowers that fade to red from late spring to fall. Attractive to many pollinators. Propagate from cuttings or seed. Dried flower heads are useful in dried flower arrangements and cutting promotes the production of more flowers.

TOP: Flowering seathrift (*Armeria maritima*) with hybrid bush lupines and tidy-tips (*Layia platyglossa*). • BOTTOM: Sulfur buckwheat (*Eriogonum umbellatum*) in full bloom.

Garden Design Note: A star of the rock garden that adds color to the late summer border.

Seaside daisy (*Erigeron glaucus*), COASTAL OR INLAND. Clumped, branched colonies of rounded, pale to blue-green leaves on stems to eight inches high. Colonies expand every year. Large, pale purple and yellow daisies in late spring and early summer. Blooms longer near the coast. Forms with white rays available. Propagate from cuttings, rooted sections, or seed. Old clumps may show dieback in the center; dividing periodically starts healthy, bushy new plants. Provide high shade and some summer water inland.
Garden Design Note: Can produce sporadic blooms twelve months a year.

Goldback fern (*Pentagramma triangularis*), COASTAL OR INLAND. [Coastal substitute: sword fern, *Polystichum munitum*]
Small, partially summer-dormant fern to around 10 inches high. Dark, polished stalks and triangular fronds covered with a gold wax on the back. Propagate from spores or divisions. Remove old brown fronds at the end of each year to promote air circulation.
Garden Design Note: Difficult to establish. Needs protection from the sun when young.

Mountain pride penstemon (*Penstemon newberryi* var. *sonomensis*), INLAND.
Matted, winter-dormant perennial with small, rounded, dark green leaves. Racemes of rose-purple to red, tubular, two-lipped flowers in summer. Hummingbird pollinated. Propagate from cuttings or seed. Short-lived in gardens. Plants should normally be restarted every few years, as they tend to become leggy and show signs of dieback.
Garden Design Note: Can be glorious with masses of magenta flowers. Loves a slope, tucked in front of a buff-colored boulder.

Hummingbird mint (*Monardella macrantha*), INLAND. [Coastal substitute: common coyote mint, *M. villosa*]
Semiprostrate, partially winter-dormant, perennial ground cover with small, highly fra-

Large flowering clone of seaside daisy (*Erigeron glaucus*).

grant, shiny, rounded leaves. Pincushionlike heads of long, tubular orange to red flowers from summer to early fall. Attractive to hummingbirds. Propagate from divisions, cuttings, or seed. Buy material in flower to determine best flower color.

Garden Design Note: Requires excellent drainage. Its large red, orange, or pink flowers make a dazzling show once established.

Lizard-tail (*Eriophyllum stachaedifolium*), COASTAL. [Inland substitute: golden yarrow, *E. confertiflorum*] Mounded subshrub or woody perennial to two or three feet high (less in wind). Elegant, deeply pinnately scooped out leaves; dark green on top, white and woolly underneath. Flat-topped cymes of small, bright yellow daisies in summer. Propagate from semihardwood cuttings or seed. To maintain a tight shape, shear the tops of the plants once a year.

Garden Design Note: A fast grower, this plant can also take the heat. Massed, it provides wonderful gray textures in a flowering border.

Succulents

Bluff lettuce (*Dudleya farinosa* and *D. caespitosa*), COASTAL.
Dense rosettes of fleshy, broadly to narrowly lance-shaped, green to gray leaves, often red tinted in the sun. Old plants may carry dozens of rosettes. Candelabralike clusters of narrow, vase-shaped, pale yellow flowers extend a few inches above leaves in early summer. Propagate from stem cuttings or seed. Old clumps sometimes show exposed stems; cut these off, shorten, and reroot to start new plants.

Garden Design Note: Tuck these in the spaces between ledge stones in a dry-stacked wall, along with sedum and strawberry.

Hot-rock dudleya (*Dudleya cymosa*), INLAND.
Rosettes of fleshy, ovate, pale green to gray green leaves. Panicles of narrow, golden to red-orange flowers in late spring. Propagate from seed. Another choice for inland sites is chalk dudleya (*D. pulverulenta*).

Garden Design Note: Looks great in low rustic bowls that feature other succulents, with gravel, rocks, and pebbles.

TOP: Flowering plant of lizard-tail (*Eriophyllum stachaedifolium*). • BOTTOM: Chalk dudleya (*Dudleya pulverulenta*) in ledgestone wall.

Common stonecrop (*Sedum spathulifolium*), COASTAL OR INLAND.
Flattened rosettes of broadly spoon-shaped green, grayish, or red-tinted leaves. Side branches send out new rosettes to colonize. Flowering stalks a few inches high carry flat-topped cymes of bright yellow, starlike flowers in midspring. Propagate from stem cuttings. The several leaf color forms make beautiful contrasts.
Garden Design Note: Mass these at a sunny sloped entry with blue-gray rocks or cobbles. Flower spikes are bright orange-yellow and rise above the rosettes dramatically.

Siskiyou lewisia (*Lewisia cotyledon*), INLAND.
Flattened rosettes of winter-dormant, broadly spoon-shaped, dark green leaves. Open cymes of showy orange, pink, rose, white, or red cactuslike flowers in spring and summer. Several named forms. Propagate from seed or offsets. Be sure to provide excellent drainage; rosettes otherwise may rot in heavy winter rains.
Garden Design Note: Difficult to establish in the ground. Try in a container first. Likes gravel.

Bulbs

Dwarf brodiaea (*Brodiaea terrestris*), COASTAL OR INLAND.
Circles of narrow grasslike leaves in early spring. Later, umbels of pale blue, starlike flowers emerge directly from the ground, reaching a few inches in height. Delightful in drifts. Propagate from cormlets or seed (three to four years to bloom). Withhold water after flowers fade or lift bulbs and store in a cool place.
Garden Design Note: Gophers love these, so be sure to plant in wire cages. Available at specialty bulb growers such as Telos Rare Bulbs.

Fawn lilies (*Erythronium californicum, E. helenae,* and *E. multiscapoideum*), INLAND.

Pairs of mottled, tongue-shaped leaves appear in late winter. In early spring, a flowering stalk carries one to several large nodding white lilylike flowers with a yellow base. Propagate from offsets or seed. Plants may be divided every few years to thin and start new clumps. Best in central and northern California.
Garden Design Note: Can be found at Telos Rare Bulbs and Far West Bulb Farm.

Sierra fawn lily (*Erythronium multiscapoideum*).

Lava and sickle-leaf onions (*Allium cratericola* and *A. falcifolium*), INLAND. [Coastal substitute: coast onion, *A. dichlamydeum*]
Miniature bulbs with two to three sickle-shaped or narrow leaves and flowering stalks four or five inches high with umbels of dark purple (*A. falcifolium*) to near-white flowers (*A. cratericola*) in early to midspring. Propagate from bulblets or seed. Withhold water after flowers fade or lift bulbs and store in a cool place.
Garden Design Note: Limited time to order these bulbs; check for availability seasonally.

Purple pussy ears (*Calochortus tolmiei*), COASTAL OR INLAND.
A single, shiny, strap-shaped leaf appears in late winter. A stalk a couple of inches high carries a few wistful, saucer-shaped, pale purple flowers covered with shaggy hairs in early to midspring. Charming effect. Propagate from bulblets or seed (four years to bloom). Withhold water after blooms fade.
Garden Design Note: Buy from specialty bulb growers.

Grasses

Tufted hair grass (*Deschampsia caespitosa* ssp. *holciformis*), COASTAL OR INLAND.
Stiff, dense tufts of pale green, narrowly lance-shaped leaves to a foot or two high. Dense panicles of flowers in summer. Shear plants back every couple of years; remove dead leaves yearly in late fall or winter.

Additional woody plants to try:

Blueblossom ceanothus (*Ceanothus thyrsiflorus*), COASTAL; Carmel creeper (*C. griseus* var. *horizontalis*), COASTAL; salal (*Gaultheria shallon*), COASTAL; beach lupine (*Lupinus chamissonis*), COASTAL; varicolored lupine (*L. variicolor*), COASTAL; hill blue bush lupine (*L. albifrons* var. *collinus*), INLAND; California sagebrush (*Artemisia californica*); bush monkeyflowers (*Mimulus aurantiacus* and *M. longiflorus*), COASTAL and INLAND; azalea-flowered monkeyflower (*M. bifidus*), INLAND; little-leaf buckwheat (*Eriogonum parvifolium*), COASTAL; Wright's buckwheat (*E. wrightii*), INLAND.

Additional perennials to try:

Douglas iris (*Iris douglasiana*), COASTAL OR INLAND; bluff angelica (*Angelica hendersonii*), COASTAL; brownie thistle (*Cirsium quercetorum*), COASTAL; yellow larkspur (*Delphinium luteum*), COASTAL; Menzies' wallflower (*Erysimum menziesii*), COASTAL; dune gumweed (*Grindelia stricta* var. *platyphylla*), COASTAL OR INLAND; dune tansy (*Tanacetum camphoratum*), COASTAL OR INLAND; California phacelia (*Phacelia californica*), COASTAL; rock phacelia (*P. imbricata*), INLAND; Brewer's rockcress (*Arabis breweri*), INLAND.

Additional succulents to try:

Dudleya attenuata and *D. lanceolata*, COASTAL; *D. edulis* and *D. arizonica*, INLAND.

Additional bulbs to try:

Blue dicks (*Dichelostemma capitatum*), COASTAL OR INLAND; dwarf form of Ithuriel's spear (*Triteleia laxa*), COASTAL; coastal form of checker lily (*Fritillaria affinis* var. *tristulis*); chocolate lily (*F. biflora*), COASTAL OR INLAND; Purdy's fritillary (*F. purdyi*), INLAND; dwarf brodiaea (*Brodiaea minor*), INLAND; star brodiaea (*B. stellaris*), INLAND; glassy brodiaea (*Triteleia lilacina*), INLAND.

Additional grasses to try:

Dune blue grass (*Poa unilateralis*), COASTAL; Nutka reed grass (*Calamagrostis nutkaensis*), COASTAL; leafy reed grass (*C. foliosa*), COASTAL; purple needlegrass (*Nassella pulchra*), COASTAL OR INLAND; California oatgrass (*Danthonia californica*), COASTAL OR INLAND.

TOP: Dwarf form coyote brush (*Baccharis pilularis* 'Twin Peaks'). • BOTTOM: Chocolate lily (*Fritillaria biflora*).

Places to Visit
Coastal Bluffs and Dunes

Patrick's Point State Park, Humboldt County. Turn west from Hwy 101 into the park about 10 miles north of Arcata. Drive to the edge of the bluffs. There are several trails that skirt the cliffs here. Look for wind-pruned forests of grand fir (*Abies grandis*), Sitka spruce (*Picea sitchensis*), coast hemlock (*Tsuga heterophylla*), native crabapple (*Malus fusca*), and wax myrtle (*Myrica californica*). Kinnikinnick forms a ground cover, and salal and coffeeberry (*Rhamnus californica*), a hedge. Perennials include false lily-of-the-valley (*Maianthemum dilatatum*), skunk cabbage (*Lysichiton americanum*), vancouveria (*Vancouveria* spp.), and several ferns.

Pt. Reyes lighthouse, Marin County. Follow Sir Francis Drake Boulevard to the last parking lot in Pt. Reyes National Seashore. You'll see steep cliffs between the parking lot and lighthouse. Look for wind-stunted evergreen huckleberry (*Vaccinium ovatum*), salal (*Gaultheria shallon*), and blueblossom ceanothus (*Ceanothus thyrsiflorus*), along with the woody ground cover kinnikinnick (*Arctostaphylos uva-ursi*). Perennials include California phacelia (*Phacelia californica*), coast wallflower (*Erysimum menziesii* var. *concinnum*), seaside daisy (*Erigeron glaucus*), lizard-tail (*Eriophyllum stachaedifolium*), coast angelica (*Angelica hendersonii*), brownie thistle (*Cirsium quercetorum*), and footsteps-to-spring (*Sanicula arc-*

topoides). Bulbs include mission bells (*Fritillaria affinis*), purple pussy ears (*Calochortus tolmiei*), coast onion (*Allium dichlamydeum*), and dwarf brodiaea (*Brodiaea terrestris*).

Pt. Lobos State Reserve, Monterey County. From

Chimney Rock near Pt. Reyes lighthouse, Marin Co.

Hwy 1, turn into the Reserve about two miles south of the Carmel River, heading south from the Monterey Peninsula. Park at Lobos Point or Whaler's Knoll and take the trails that follow the cliffs. The forest consists of Monterey pine (*Pinus radiata*), Monterey cypress (*Cupressus macrocarpa*), and coast live oak (*Quercus agrifolia*). Perennials and small shrubs include bluff-lettuce (*Dudleya farinosa*), lizard-tail, woolly aster (*Lessingia filaginifolia* var. *californica*), dune sagebrush (*Artemisia pycnocephala*), California sagebrush (*A. californica*), deerbroom lotus (*Lotus scoparius*), sticky monkeyflower (*Mimulus aurantiacus*), coast plantain (*Plantago maritima*), Indian paintbrush (*Castilleja wightii*), and seaside daisy.

Nipomo Dunes, San Luis Obispo County. In Santa Maria, turn west from Hwy 101 onto Hwy 166. On reaching Hwy 1, turn north and go a few miles, then, following signs to Oso Flaco Lake, continue west a couple more miles. The dunes are in a Nature Conservancy preserve and are home to many plants adapted to bluff and rock-garden conditions. Look for wind-pruned hedges of wax myrtle and arroyo willow (*Salix lasiolepis*). Perennials include coast sand-verbena (*Abronia maritima*), pink sand-verbena (*A. umbellata*), Nipomo Dunes coyote mint (*Monardella crispa*), little-leaf buckwheat (*Eriogonum parvifolium*), Blochman's bush senecio (*Senecio blochmanae*), seaside daisy, yarrow (*Achillea millefolium*), dune sagebrush, lizard-tail, Santa Barbara wallflower (*Erysimum suffrutescens* var. *lompocense*), and cobweb thistle (*Cirsium occidentale*).

Prisoner's Harbor, Santa Cruz Island, Santa Barbara County. Boat trips can be arranged from the Ventura harbor to this put-in point along the north side of Santa Cruz Island. Plants include Greene's dudleya (*Dudleya greenei*), California fuchsia (*Epilobium canum*), gray-leafed goldenbush (*Hazardia detonsa*), Santa Cruz Island buckwheat (*Eriogonum arborescens*), island bush monkeyflower (*Mimulus flemingii*), and chicory-leafed wandflower (*Stephanomeria cichoriacea*).

Inland Cliffs

Table Mountain, near Oroville, Butte County. Take Cherokee Road from the town of Oroville (northern Sierra foothills). The flat top of Table Mountain consists of columnar basalt. Rock ferns include birdsfoot fern and goldback fern. Annuals include annual stonecrop (*Parvisedum pumilum*) and two kinds of jewelflower, *Streptanthus diversifolius* and *S. tortuosus* var. *suffrutescens*. Perennials include padre's shooting star (*Dodecatheon clevelandii*), bitterroot, oak violet (*Viola purpurea* var. *quercetorum*), spikemoss (*Selaginella bigelovii*), and California pipevine (*Aristolochia californica*). Bulbs include glassy brodiaea (*Triteleia lilacina*), golden brodiaea (*T. ixioides*), odontostomum (*Odontostomum hartwegii*), mariposa tulips (*Calochortus* spp.), lava onion (*Allium cratericola*), and sickle-leaf onion.

Magalia serpentine barrens, Butte County. Take the Skyway from Chico through Paradise east to the signal in Magalia. Park just east on Coutelanc Road and walk into the open areas of bluish serpentine rock. Trees include California bay, gray pine (*Pinus sabiniana*), and Macnab cypress (*Cupressus macnabiana*). Shrubs include yerba santa (*Eriodictyon californicum*), bush poppy (*Dendromecon rigida*), and buckbrush (*Ceanothus cuneatus*). Perennials include azure penstemon (*Penstemon azureus*), sulfur buckwheat, needlegrass (*Nassella* sp.), Lewis Rose's butterwort (*Senecio lewisrosei*), and woolly daisy (*Eriophyllum lanatum*). Bulbs include Sierra fawn lily (*Erythronium multiscapoideum*), blue dicks, purple pussy ears, and

California brodiaea. Ferns include goldback fern and Indian's dream (*Aspidotis densa*).

The palisades near Mt. St. Helena, Napa County. Hike up Oathill Mine Road from Calistoga (at the head of the Napa Valley) or walk down from Robert Louis Stevenson State Park, where Hwy 29 crosses over a hump on Mt. St. Helena. Either way, you need to walk several miles to access the heart of this rugged terrain, featuring pinnacles of volcanic tuff. Rock ferns include goldback fern (*Pentagramma triangularis*), birdsfoot fern (*Pellaea mucronata*), coffee fern (*P. andromedifolia*), and California lace fern (*Aspidotis californica*). Perennials include bitterroot (*Lewisia rediviva*), shooting stars (*Dodecatheon hendersonii*), red rock penstemon (*Keckiella corymbosa*), redberry buckthorn (*Rhamnus crocea*), and scarlet larkspur (*Delphinium nudicaule*). Bulbs include Purdy's fritillary (*Fritillaria purdyi*), sickle-leaf onion (*Allium falcifolium*), California brodiaea (*Brodiaea californica* var. *leptandra*), and blue dicks (*Dichelostemma capitatum*).

The top of Mt. Diablo, Contra Costa County. Take the Summit Trail, which departs from the parking lot just below the summit museum and makes a one-mile loop. Trees here include California bay (*Umbellularia californica*), interior live oak (*Quercus wislizenii*), and canyon live oak (*Q. chrysolepis*). Perennials include scarlet larkspur, rayless arnica (*Arnica discoidea*), bitterroot, red rock penstemon, blue foothill penstemon (*Penstemon heterophyllus*), bristly phacelia (*Phacelia nemoralis*), linear-leaved goldenbush (*Ericameria linearifolia*), and sulfur buckwheat (*Eriogonum umbellatum*). Bulbs include sickleleaf onion, Howell's onion (*Allium acuminatum*), and Venus mariposa tulip (*Calochortus venustus*).

Figueroa Mountain, Santa Barbara County. From the town of Los Olivos in the Santa Ynez Valley, take Figueroa Road up the mountain. There are several areas of serpentine and other rock outcrops with attractive plants. Look for coast live oak, canyon live oak, sycamore, California bay, and gray pine. Wildflowers include phacelias, shooting stars, foothill wallflower (*Erysimum capitatum*), linanthuses (*Linanthus* spp.), golden pincushions (*Chaenactis glabriuscula*), California poppy (*Eschscholzia californica*), and goldfields (*Lasthenia californica*). Perennials include long-stemmed buckwheat (*Eriogonum elongatum*), wild pansy (*Viola pedunculata*), and rock phacelia (*Phacelia imbricata*). Bulbs include red-skinned onion (*Allium haematochiton*), Catalina mariposa (*Calochortus catalinae*), white globe tulip (*C. albus*), blue dicks, and chocolate lily (*Fritillaria biflora*).

CHAPTER 2

REDWOOD FOREST: GARDENING UNDER COOL GIANTS

The towering trees of *Sequoia sempervirens*—coast redwood—create hushed, cathedral-like aisles in California's fog belt from southern Monterey County to just north of the Oregon border. Conditions that favor redwoods include mild winters with little frost; cool, moist, foggy summers; lack of salt-laden coastal winds; and deep, silty, river-deposited soils. The climate, together with the decomposition of redwood needles, creates decidedly acid soils with a pH of 5.0 to 5.5. The plants associated with redwood forests flourish best under these conditions and are excellent choices for coastal gardens in the fog belt.

A mature redwood forest has a simple structure: a dense canopy created by stands of very tall trees up to 350 feet high; an understory filled with ferns, herbaceous ground covers, and seasonal wildflowers; and borders—along streams or at the forest's edge—

of dense, nearly impenetrable seasonal shrubs and small trees. The lushest redwood forests are found in Humboldt and Del Norte counties.

Much of California's original redwood forest has been degraded by logging. Although redwoods grow rapidly in their first few decades, second-growth forests may take hundreds of years to achieve the stability, grandeur, and diversity of the original forests. Many other kinds of trees enter into the mix of a secondary forest, including California bay (*Umbellularia californica*), madrone (*Arbutus menziesii*), tanbark oak (*Lithocarpus densiflorus*), grand fir (*Abies grandis*), coast hemlock (*Tsuga heterophylla*), Sitka spruce (*Picea sitchensis*), and Douglas-fir (*Pseudotsuga menziesii*). These trees are good associates for redwoods on large estates with ample room.

LEFT: Redwood trees (*Sequoia sempervirens*). • ABOVE: Waterfall in redwood forest, Jedediah Smith State Park.

Bench (not visible)

Hedge across the street

Grove of dogwoods

Street

Existing turf

Driveway

Redwood duff footpath

Hillsborough boulders

Creating a Garden

You can create a miniature redwood forest on a very small scale by establishing a shaded container garden with redwood associates. Or you can use a single redwood tree or other moisture-loving conifer—Monterey pine or bishop pine (*Pinus radiata* and *P. muricata*), Sitka spruce (*Picea sitchensis*), western hemlock (*Tsuga heterophylla*), grand fir (*Abies grandis*), Port Orford cedar (*Chamaecyparis lawsoniana*), or Douglas-fir (*Pseudotsuga menziesii*). But bear in mind that these trees require ample space and have shallow, wide-spreading, thirsty roots, so underplantings need some supplemental summer water even in the fog belt. Be sure to position plants according to their light requirements—full deep shade, light peripheral shade, shade part of the day, and other variations.

For a natural design, utilize natural groupings of shrubs, interwoven with seasonal and evergreen ground covers and swaths of ferns and seasonal wildflowers. Bloom time starts in February with slink pods (*Scoliopus bigelovii*) and milkmaids (*Cardamine californica*); peaks in April; and generally finishes by June with bead-lily (*Clintonia andrewsiana*), twisted-stalk (*Streptopus amplexifolius*), and firecracker brodiaea

Rendering of featured garden.

(*Dichelostemma ida-maia*). Additional seasonal drama is provided by the unfurling of fern fiddleheads in early to midspring, the ripening of berries in summer and fall, and the flush of fall color in deciduous shrubs and herbs. Careful planning of the plant palette helps assure something of interest throughout the year.

Design Notes

The house is an Italian Renaissance design built in 1926. It is ficus vine-covered and surrounded by mature trees, including a redwood grove, Monterey cypress,

CLOCKWISE FROM TOP LEFT: California wax myrtle (*Myrica californica*) and flowering dogwood (*Cornus nuttallii* 'Eddie's White Wonder') in the featured garden. • Leaves and flowers of western azalea (*Rhododendron occidentale*) on the rear deck of the featured garden. Both photographs by S. Ingram. • Redwood community painting by A. Yankellow.

deodar cedar, Monterey pines, and Japanese maples. The gardens reflect a traditional estate garden style, now reinterpreted using California native plantings. Because the garden derives moisture from fog and has mature trees on site, natives from the redwood community as well as other conifers that occur in our coastal ranges made ideal additions. Ample shade is available in the rear garden, where we have restored stone terraced planters and created a species garden of hydrangeas and viburnums; these are then underplanted with ground covers native to the redwood community.

Adjacent to the redwood garden is a a dwarf-conifer rock garden. Each dwarf tree is a selection of a native California conifer, though most were acquired in Oregon from specialty nurseries. There are literally hundreds of choices of dwarf California conifers, of which redwoods and Douglas-firs are especially attractive and useful. My cli-

TOP: Redwood garden with plantings of dwarf conifers and bearberry adjacent to the lawn. Photograph by S. Ingram.
• BOTTOM: Back patio tiles painted to depict wild ginger. Photograph by S. Ingram.

ent, who is originally from Michigan, grew up with dwarf-conifer foundation shrubs around his home. He liked the conifer garden so much, we decided to add another one on the other side of the driveway.

We created a new front entry and rear patio, taking as our inspiration the plants and flowers of the redwood forests. Ananda Yankellow, our artist-in-residence, hand-painted rustic ceramic tiles with designs of redwood, western azalea, wild ginger, bleeding heart, huckleberry, redwood sorrel, western sword fern, and dogwood. On the rear patio, she glazed tiles to create a redwood landscape with the vine-covered home at the center of the composition. In each of our gardens, we try to introduce original art inspired by the natural beauty of California; we particularly like to use motifs that link the site to the surrounding natural plant community. In this case, the giant redwood trees that were planted nearly eighty years ago provided the inspiration for both the art and the garden.

Scope of Work

- ◆ Create a garden path of redwood duff or fine-textured, aged redwood mulch.
- ◆ Install stone boulders for accent and rock walls.
- ◆ Create berms between boulders and between pathways.
- ◆ Install antique stone bench between two redwood trees, flanked by dogwoods.
- ◆ Install plantings. Accent berms with plantings.
- ◆ Retrofit irrigation.
- ◆ Mulch planting beds with fine decomposed redwood mulch.

TOP: Front patio tiles, some painted with coast redwood motif. Photograph by S. Ingram. • BOTTOM: Dwarf redwood (*Sequoia sempervirens* 'Albo-Spica'). Photograph by S. Ingram.

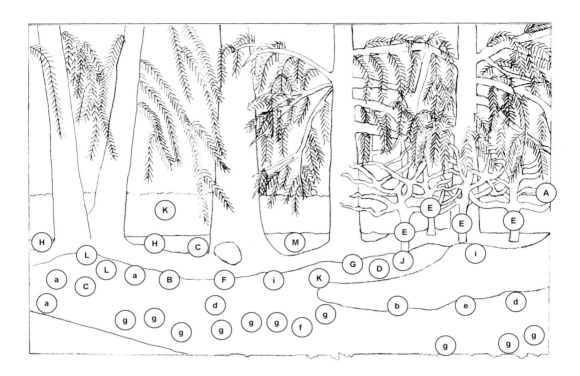

Plant List

Symbol	Botanical Name	Common Name
A	*Acer circinatum*	vine maple
B	*Achlys triphylla*	vanilla leaf
C	*Adiantum aleuticum*	five-finger fern
D	*Asarum caudatum*	wild ginger
E	*Cornus nuttallii* 'Eddie's White Wonder'	flowering dogwood
F	*Ceanothus griseus* var. *horizontalis* 'Yankee Point'	California lilac
G	*Lysichiton americanum*	skunk cabbage
H	*Oxalis oregana*	redwood sorrel
i	*Polystichum munitum*	western sword fern
J	*Rhododendron macrophyllum*	rosebay rhododendron
K	*R. occidentale*	western azalea
L	*Vaccinium ovatum*	evergreen huckleberry
M	*Vancouveria hexandra*	inside-out flower

Featured garden key.

Dwarf Conifer Garden

a	*Rhamnus californica* 'Seaview'	coffeeberry
not shown	*Pseudotsuga menziesii* 'Graceful Grace'	dwarf Douglas-fir
b	*P. menziesii* 'Fletcheri'	Douglas-fir
not shown	*P. menziesii* 'Loggerhead'	Douglas-fir
c	*Pinus contorta* 'Spaans Dwarf'	dwarf shore pine
d	*Sequoia sempervirens* 'Adpressa'	dwarf coast redwood
e	*S. sempervirens* 'Prostrata'	prostrate coast redwood
not shown	*S. sempervirens* 'Kelly's Prostrate'	Kelly's dwarf redwood
not shown	*Abies lasiocarpa* 'Green Globe'	dwarf subalpine fir
not shown	*Thuja plicata* 'Grune Kugel'	dwarf western red cedar
not shown	*Chamaecyparis lawsoniana* 'Nova'	dwarf Lawson's cypress
f	*C. lawsoniana* 'Gimbornii'	Lawson's cypress
g	*Arctostaphylos uva-ursi* 'Pt. Reyes'	kinnikinnick
not shown	*Ceanothus maritimus*	maritime ceanothus
not shown	*Erigeron glaucus*	seaside daisy

Constructed waterfall with redwood plantings.

Plants to Use

Small Trees

Vine maple (*Acer circinatum*).
Slow-growing, deciduous tree to 25 feet high. Tiers of horizontally spreading branches,

multilobed leaves (similar to Japanese maple), small clusters of maroon-and-white flowers in early spring, pink samaras (winged fruits) in summer, and flame-red leaves in autumn. Favors the edge of redwood forests in the north. Propagate from doubly stratified seed. Trees may occasionally need a selective pruning to remove overly dense branches. Leaves make an excellent mulch.

Garden Design Note: Excellent with wild grape, and with dogwood for fall color. Substitute for Japanese maples in an Asian-themed native garden.

California hazelnut (*Corylus cornuta* var. *californica*).

Fast-growing deciduous tree to 15 feet tall. Tiers of horizontal branches; soft, downy, broadly elliptical, toothed leaves; fuzzy-husked edible nuts ripening in late summer; and graceful, pendant male catkins from late winter to early spring. Favors the edge of redwood, mixed-evergreen, and ponderosa pine forests. Propagate from suckers or fresh seed. Plants need little extra care except occasional pruning to shape.

TOP: Leaves turning color on vine maple (*Acer circinatum*). • BOTTOM: California hazelnut (*Corylus cornuta* var. *californica*).

Garden Design Note: You can carefully shape this small tree for an Asian effect. Curved branches that display male catkins against a stark wall in the spring are spellbinding.

California wax myrtle (*Myrica californica*).
Fast-growing evergreen shrub or tree to 20 feet tall. Can be pruned into a dense hedge or maintained as a single-trunked specimen. Features tight, upward-trending branches thickly clothed in long, glossy, subtly toothed, spicily fragrant, lance-shaped leaves; and tiny axillary clusters of greenish male or female flowers that ripen into warty, dark purple berries in summer. Occurs mostly on the edge of conifer forests in the fog belt from south-central California northward. Propagate from semihardwood cuttings. Shear new growth as needed to shape the plants and keep branches dense and full. Garden Design Note: Wonderful dark green screening plant. With irrigation it can grow twelve feet in three years. A small-leaved selection called 'Buxifolia' shears well.

Large Shrubs

Western azalea (*Rhododendron occidentale*).
Multibranched deciduous shrub to 10 feet tall with twigs in umbrellalike clusters. Nar-

Privacy screen with California wax myrtle (*Myrica californica*).

rowly elliptical, pale green leaves are skunk scented on hot days; leaves turn bronze to orange in fall. Dense clusters of large white to deep rose, sweetly fragrant flowers appear in late spring or early summer; each blossom has a yellow or orange patch on the upper petals. Fuse-shaped seed pods contain fine, dustlike seed. Follows permanent streams. Propagate from hardwood cuttings (bottom heat), layered

branches, or seed sprinkled on finely milled peat moss. Multitrunked specimens may need some thinning of canes from time to time to keep the overall structure open. Garden Design Note: Use at the edge of a shade canopy. Needs sun for profuse blooms. Mass for effect. Powerful fragrance.

Rosebay rhododendron (*Rhododendron macrophyllum*).
Large, multibranched shrub or small tree to 15 feet high with large, tough, elliptical, evergreen leaves. Clusters of showy, but not fragrant, rose-purple flowers in late spring.

TOP: Leaves and flowers of western azalea (*Rhododendron occidentale*). • BOTTOM: Flowers of rosebay rhododendron (*Rhododendron macrophyllum*).

Fuse-shaped seed pods have tiny, dustlike seed. Thrives on the edge of northern redwood and other coastal forests. Blooms best with some sun. Propagate from hardwood cuttings (bottom heat) or seed. Little maintenance is needed; however, plants are slow to establish and do best with a fungal innoculum.

Garden Design Note: Hard to find in California. Oregon has cultivars, but they are difficult to establish. Use organic mulch.

Evergreen huckleberry (*Vaccinium ovatum*).
Multibranched evergreen shrub to 12 feet high with moderate to slow growth. Branches with two rows of small, narrowly lance-shaped, glossy leaves edged with fine teeth. White to pale pink, bell-shaped flowers under branches from late winter to midspring.

Redwood garden with redwood sorrel (*Oxalis oregana*), western sword fern (*Polystichum munitum*), and evergreen huckleberry (*Vaccinium ovatum*). Photograph by S. Holt.

Luscious, blueberry-flavored fruits in late summer. Favors coastal coniferous forests. Propagate from hardwood cuttings or stratified seed.

Garden Design Note: Very slow, but worth the effort. Huckleberry jam is the best. Plant it when you have a baby; then you can have jam when she is a teenager.

Salmonberry (*Rubus spectabilis*).

Fast-growing, multitrunked deciduous shrub to 15 feet high. Trifoliate, pinnately veined leaves; branches lined with prickles. Rose-red, roselike flowers in spring; salmon egg-like berries in summer. Prefers edge of coastal coniferous forests. Propagate from suckers or root divisions. Prune heavily when growth becomes intertwined to maintain an open structure and more pleasing balance of branches, and remove excess suckers that sprout from the base or from the roots. May be invasive.

Garden Design Note: Beautiful pink, strawberrylike flower looks good with thimbleberry on a slope. Many selections are available from Oregon nurseries.

Small Shrubs

Thimbleberry (*Rubus parviflorus*).

Fast-growing, potentially invasive, deciduous shrub to six feet tall. Shreddy brown bark; large maple-shaped, softly textured leaves turning pale gold in fall; clusters of large white single roselike blossoms in mid to late spring. Thimble-shaped dark red berries have a delicious raspberry flavor. Prefers forest edges near streams. Propagate from root divisions. Thin extra branches once a year to avoid an overly dense structure, and watch for excessive suckering from roots.

Garden Design Note: Intersperse with western sword fern and western columbine along a shady creek bed.

Long-leaf barberry (*Berberis* [= *Mahonia*] *nervosa*).

Slow-to-establish, colonizing, evergreen shrub to two or three feet high. Long, pinnately compound leaves with leaflets that resemble glossy holly leaves at the end of canelike stems; yellow flowers in clustered slender spikes in late spring. Bowling pin-shaped pale purple berries are sour in taste. Favors moist coniferous forests in the fog belt. Propagate by root division or stratified seed.

Flowers and fruits of long-leaf barberry (*Berberis nervosa*).

Garden Design Note: If you have the patience, this striking, deep green, low-growing shrub will reward you with year-round lushness. Plant with faster-growing grasses and ground covers.

Red huckleberry (*Vaccinium parvifolium*).
Deciduous shrub with many slender broomlike branches to six feet high. Angled green twigs with small, narrowly oval leaves. Tiny, greenish or pinkish, lanternlike flowers hide under branch tips in early to midspring. Scarlet berries of excellent flavor appear in early summer. Frequently begins life on old logs and nurse stumps and grows quickly. Mostly confined to coniferous forests on the north coast. Propagate from divisions of older plants or from stratified seed.
Garden Design Note: Performs best with a thick organic layer of mulch or conifer duff.

Salal (*Gaultheria shallon*).
Dense, wandering, evergreen shrub to six feet high. Zigzag stems and large, leathery, ovate leaves; lines of white or pale pink, lanternlike flowers on horizontal stems in mid to late spring; fat, near-black berries of good flavor. Mostly confined to coastal coniferous forests from central California north. Propagate from divisions. Be sure that plants don't become stressed during the dry season; they're vulnerable to attack by thrips. Large colonies may also need periodic thinning.

Garden Design Note: Needs extra water, but grows under conifers. Use thick organic mulch to assist in establishment. Berries are delicious. Great foliage for bouquets.

Ferns

Propagation Note: Ferns can be divided when they develop more than one crown, or started from spores using a sterile technique.

Western sword fern (*Polystichum munitum*).
Tufted evergreen fern two to four feet tall. Leathery fronds are once pinnately compound, with each segment resembling a

Clumps of western sword fern (*Polystichum munitum*).

73

miniature sword. New fiddleheads look like fuzzy caterpillars. Typifies the understory of dark coniferous woods along California's coast. Remove old brown fronds each winter.

Garden Design Note: Mass this fern under redwoods with additional irrigation. This is the most universal fern for wet, shady gardens. Can also handle dry shade, but won't get as large.

Spiny wood fern (*Dryopteris expansa*).
Tufted, semievergreen fern three feet high. Fronds are almost three times pinnately divided into lacy segments, with each frond describing a broad triangle. Favors moist coniferous woods on California's north coast. Remove old brown fronds each winter.
Garden Design Note: Mass on a shady slope. Mix with evergreen ferns.

Five-finger fern (*Adiantum aleuticum*).
Small, winter-deciduous fern, seldom exceeding a foot in height. Each shiny black stipe

Driveway with concrete-block retaining wall studded with inside-out flower (*Vancouveria planipetala*), western sword fern (*Polystichum munitum*), five-finger fern (*Adiantum aleuticum*), and common alumroot (*Heuchera micrantha*). Photograph by S. Holt.

is forked, then redivided into five or more narrow, fingerlike segments; the ultimate segments are broadly
crescent shaped. An especially delicate fern that is
beautiful massed. Occurs by seeps, waterfalls, and
stream courses. Cut off old fronds as new ones appear
in spring.
Garden Design Note: Place this delicate maidenhair in
the shade near flowing water or a fountain that splashes
on it.

Deer fern (*Blechnum spicant*).
Small, tufted, semi-winter-dormant fern, rarely exceeding a foot in height. Splayed sterile fronds are once
pinnately divided, with a narrowly bowed outline and
stiff, erect fertile fronds. New fiddleheads are elegantly
striped. Grows in wet ditches, seeps, and streamsides.
Garden Design Note: Likes more water than some of
the other ferns. Display in an area where the fertile
fronds, a curious feature of the deer fern, are visible.

Ground Covers

Propagation Note: Most ground covers are easily
propagated by division of creeping stems or roots.

Wild ginger (*Asarum caudatum*).
Evergreen, ginger-scented creeper with large, handsome, dark green, broadly heart-shaped leaves. Bizarre,
tailed maroon blossoms hide under leaves. Grows in
the dense shade of coniferous forests.
Garden Design Note: Can be easier to establish dur-

ing the warm months. Buy well-developed plants and
you will be rewarded quickly with a lush green
ground cover.

Redwood sorrel (*Oxalis oregana*).
Evergreen, stoloniferous plant with trifoliate, clover-
like leaves, each leaflet an upside-down heart. Leaflets often bear a whitish stripe down the middle

TOP RIGHT: Understory of deer fern (*Blechnum spicant*, foreground) and lady ferns (*Athyrium filix-femina*). MIDDLE RIGHT:
Ground cover of wild ginger (*Asarum caudatum*). • LEFT: Redwood sorrel (*Oxalis oregana*).

and are purple underneath. Single white, pink, or rose-purple blossoms appear above leaves in early to midspring. Occurs chiefly under redwoods.

Garden Design Note: Easy to establish with adequate irrigation under redwoods. Mix with wild ginger for a rich texture.

Vanilla leaf (*Achlys triphylla*).

Winter-dormant, rhizomatous perennial with large, trifoliate, pale green leaves. Each leaflet resembles a fanciful butterfly wing. Narrow spikes of white flowers in late spring or early summer. Leaves sweet scented when dry. Carpets north coastal coniferous forests.

Garden Design Note: Though difficult to find in California, it is available at Oregon nurseries. Mass plant one foot on center in soils rich in humus or organic matter. Slow to establish.

Inside-out flower (*Vancouveria hexandra*).

Coarse, rhizomatous ground cover to 10 inches high. Pinnately compound leaves feature ivy-shaped leaflets, resembling the leaflets of their close Asian relatives, the epimediums. Leaves die back in winter. Open racemes or panicles of small white blossoms with recurved petals appear in midspring. Thrives in coniferous forests. Propagate from divisions or seed. Cut off old growth when new shoots appear in spring.

Garden Design Note: This very tidy, small-leaved ground cover can be appropriate in small spaces. Mix with evergreen inside-out flower (*V. planipetala*).

Redwood violet (*Viola sempervirens*).

Ground-hugging, evergreen trailer, rooting as it grows. Leaves are dark green and nearly

round. Single-stemmed yellow violets appear sporadically but peak in April. Prefers banks in coastal coniferous forests.

Garden Design Note: Difficult to find.

Western bleeding heart
(*Dicentra formosa*).

Wandering, winter-dormant, rhizomatous ground cover, invasive with summer water. Coarsely divided, pale green, fernlike leaves; small clusters of white, pale pink, or

Western bleeding heart (*Dicentra formosa*). Photograph by S. Ingram.

dark rose-purple, heart-shaped flowers in mid to late spring. Leaves die back early without water. Found throughout the redwood region.

Garden Design Note: Takes a few years to establish. Mix with other, faster-growing redwood ground covers like sorrel and wild ginger.

Sugar scoop (*Tiarella trifoliata* var. *unifoliata*).

Densely knit colonies are winter dormant. Each round leaf is shallowly lobed and subtly scalloped. Slender racemes of small white, nodding, bell-shaped flowers appear in late spring and summer. Favors coastal coniferous forests. Propagate by divisions.

Garden Design Note: Most effective if massed, as the flower is tiny. Plant with other small-textured plants, like inside-out flower and redwood violet. Best in small-scale, cool, damp gardens.

False lily-of-the-valley (*Maianthemum dilatatum*).

Stoloniferous, winter-dormant ground cover with broad, rounded leaves inscribed with parallel veins. Narrow spikes of tiny white flowers appear in May. Grows throughout north coastal forests. Propagate by division.

Garden Design Note: Takes a few years to establish, and rich organic matter, adequate water, and the shade of conifers are necessary. Difficult to establish in summer heat.

Wildflowers

Red columbine (*Aquilegia formosa*).

Winter-dormant perennial two to three feet high in flower. Coarsely compound, bluish green leaves. Hanging red and yellow flowers with long nectar spurs occur from midspring into summer. Requires well-drained soil, light to moderate shade, and occasional water. Propagate from seed. Remove the old flowering stalks and leaves in winter.

Garden Design Note: Mass this hummingbird plant along a water course or near a fountain that splashes. Plant with meadow rue, a close relative.

Smith's fairy bells (*Disporum smithii*).

Dense colonies of branched, 18-inch-tall stems with glossy, ovate leaves. Slender, white, bell-shaped flowers cluster under branch tips in midspring, with bright orange berries appearing in summer. Abundant in northern redwood forests. Propagate from divisions or stratified seed.

Garden Design Note: Colonies of fairy bells are spectacular on a shady bank above a rock wall.

Red columbine (*Aquilegia formosa*).

Trilliums (*Trillium ovatum* and *T. chloropetalum*).
Slow growth into colonies from deep-seated tubers. Three broadly ovate leaves offset a single showy blossom. *T. ovatum* (wakerobin) carries its fragrant flower on a slender stalk above the leaves; flowers open white and gradually fade to deep rose-purple. *T.*

chloropetalum (giant trillium) holds its pure white, pink, greenish yellow, or maroon-red flowers directly in the center of the leaves. Propagate by divisions of colonies, since seed takes up to seven years to bloom.

Garden Design Note: You may find these memorable beauties at native plant sales and specialty nurseries in Oregon.

Bead lily (*Clintonia andrewsiana*).
Two to four glossy, broadly elliptical leaves emerge from a cluster of deep-seated fleshy roots in early spring. Panicles of pink-purple flowers appear at spring's end, with waxy blue, beadlike berries following in early summer. Restricted to coastal coniferous forests. Propagate from root divisions or seed (three to four years to reach blooming size). Deer are fond of the flowers and fruits.
Garden Design Note: This plant is difficult to find but well worth the effort.

False Solomon's seal (*Smilacina racemosa*).
From a plump "asparagus spear" in early spring comes a stout stalk with broadly ovate leaves and a dense terminal panicle of numerous tiny cream-colored flowers with lily-of-the-valley fragrance, with purple-speckled berries appearing in summer. Grows into multistemmed clumps. Widespread in coniferous and mixed-evergreen forests. Propagate from divisions or seed (around three years to bloom from seed).
Garden Design Note: Plant several in a cluster behind a bench on a shady slope. The fragrance will intoxicate. Mix with ferns and iris.

Flowering plant of wakerobin (*Trillium ovatum*).

Additional small trees to try:

Flowering dogwood (*Cornus nuttallii*).

Additional large shrubs to try:

Cascara sagrada (*Rhamnus purshiana*), mock azalea (*Menziesia ferruginea*), stink currant (*Ribes bracteosum*), pink-flowering currant (*R. sanguineum* var. *glutinosum*).

Additional small shrubs to try:

Western burning bush (*Euonymus occidentalis*), twinberry honeysuckle (*Lonicera involucrata*), wood rose (*Rosa gymnocarpa*), canyon gooseberry (*Ribes menziesii*).

Additional ferns to try:

Lady fern (*Athyrium filix-femina*), giant chain fern (*Woodwardia fimbriata*), licorice fern (*Polypodium glycyrrhiza*), leather fern (*P. scouleri*).

Additional ground covers to try:

Goldthread (*Coptis laciniata*), inside-out flower (*Vancouveria planipetala*), redwood waterleaf (*Hydrophyllum tenuipes*), forest anemone (*Anemone deltoidea*), twinflower (*Linnaea borealis*), bunchberry (*Cornus canadensis*), piggyback plant (*Tolmiea menziesii*), western coltsfoot (*Petasites frigidus* var. *palmatus*).

Additional wildflowers to try:

Baneberry (*Actaea arguta*), slink pod (*Scoliopus bigelovii*), Hooker's fairy bells (*Disporum hookeri*), starry Solomon's plume (*Smilacina stellata*), twisted-stalk (*Streptopus amplexifolius*), western heartsease (*Viola ocellata*), smooth yellow violet (*V. glabella*), fringe-cups (*Tellima grandiflora*), rue-anemone (*Anemone oregana*), fairy slipper (*Calypso bulbosa*), milkmaids (*Cardamine californica*), California harebell (*Campanula prenanthoides*), California milkwort (*Polygala californica*), toothed monkeyflower (*Mimulus dentatus*).

Container with dwarf redwood (*Sequoia sempervirens* 'Kelley's Prostrate'), western bleeding heart (*Dicentra formosa*), and the varigated form of piggyback plant (*Tolmiea menziesii* 'Variegata'). Photograph by S. Ingram.

Places to Visit

Jedediah Smith State Park, Del Norte County.
This park lies north of Crescent City on Hwy
199 just beyond its intersection with Hwy 101.
(There is at least one other entrance.) Here you
can see some of the most majestic redwood forest, where huge trees and lush understory are
fed by heavy winter rains and constant summer fogs. Shrubs include California hazelnut
(*Corylus cornuta* var. *californica*), red and evergreen huckleberries (*Vaccinium parvifolium* and
V. ovatum), salal (*Gaultheria shallon*), western
azalea (*Rhododendron occidentale*), rosebay rhododendron (*R. macrophyllum*), and vine maple
(*Acer circinatum*). Ferns include sword fern
(*Polystichum munitum*), five-finger fern (*Adiantum aleuticum*), deer fern (*Blechnum spicant*),
and lady fern (*Athyrium filix-femina*). Herbaceous plants include fairy bells (*Disporum* spp.),
false Solomon's seals (*Smilacina* spp.), coast
trillium (*Trillium ovatum*), smooth yellow violet
(*Viola glabella*), baneberry (*Actaea arguta*),
skunk cabbage (*Lysichiton americanum*), redwood sorrel (*Oxalis oregana*), and wild ginger
(*Asarum caudatum*).

Prairie Creek State Park, Humboldt County. This
park is located about 35 miles north of Eureka
and just a few miles north of Orick (headquarters for Redwood National Park). The park
abuts Hwy 101; you can also drive out to the
coast on Gold Beach Road. Trails within the park
lead from pure redwood forest into north
coastal coniferous forest, sand dunes, bogs, and
magnificent Fern Canyon. Many of the species
here are the same as in Jedediah Smith, but
there is much admixture of other trees such
as Lawson cypress (*Chamaecyparis lawsoniana*), coast hemlock (*Tsuga heterophylla*), Sitka
spruce (*Picea sitchensis*), grand fir (*Abies
grandis*), and red alder (*Alnus rubra*). Highlights
near the coast include toothed monkeyflower
(*Mimulus dentatus*), leather fern (*Polypodium
scouleri*), twisted-stalk (*Streptopus amplexifolius*), and stink currant (*Ribes bracteosum*).
Unusual shrubs include mock azalea (*Menziesia
ferruginea*) and cascara sagrada (*Rhamnus purshiana*).

Armstrong State Redwood Reserve, Sonoma

Ferns and shrubs in front of tall redwoods (*Sequoia sempervirens*), Jedediah Smith State Park, Del Norte County.

County. Armstrong Redwoods is located about three miles north of Guerneville and the Russian River. From River Road, turn north on Armstrong Road to the park. Several trails loop through the main canyon and over the ridges. The floodplain is home to some large virgin redwoods, while the adjacent ridges are redwood forest that gradually merges into mixed-evergreen forest. Armstrong Redwoods is a nice example of redwood forest vegetation in the central part of its range. Other trees include Douglas-fir (*Pseudotsuga menziesii*), California bay (*Umbellularia californica*), madrone (*Arbutus menziesii*), tanbark oak (*Lithocarpus densiflorus*), and bigleaf maple (*Acer macrophyllum*). Shrubs include evergreen huckleberry, blackcap raspberry (*Rubus leucodermis*), California hazelnut, wood rose (*Rosa gymnocarpa*), blueblossom ceanothus (*Ceanothus thyrsiflorus*), cream bush (*Holodiscus discolor*), and thimbleberry (*Rubus parviflorus*). Herbaceous plants include slink pod (*Scoliopus bigelovii*), bead-lily (*Clintonia andrewsiana*), calypso orchid (*Calypso bulbosa*), redwood sorrel, wild ginger, trail plant (*Adenocaulon bicolor*), redwood violet (*Viola sempervirens*), fairy bells, false Solomon's seal, coast trillium, and milkmaids (*Cardamine californica*). Ferns include sword fern, common wood fern (*Dryopteris arguta*), Dudley's shield fern (*Polystichum dudleyi*), goldback fern (*Pentagramma triangularis*), and bracken (*Pteridium aquilinum*).

Purissima Canyon, San Mateo County. Purissima Canyon is located on the San Mateo coast. Go four miles south from Half Moon Bay on Hwy 1, turn left onto Verde Road, then stay straight—Verde Road makes a bend to the right, but by continuing straight you are following Purissima Road, which takes you to a parking lot at the mouth of the canyon. You can walk up the canyon through redwood forest or make a loop through redwoods, mixed-evergreen forest, and north coastal scrub. The second-growth redwoods here feature a lush understory of evergreen huckleberry, creambush, western burning bush (*Euonymus occidentalis*), California wax myrtle (*Myrica californica*), and creek dogwood (*Cornus sericea*) along with perennials such as fairy bells, false Solomon's seal, western coltsfoot (*Petasites frigidus* var. *palmatus*), trail plant, star flower (*Trientalis latifolia*), trillium, bead-lily, and great masses of sword, lady, and five-finger ferns.

Note: Redwood forests south of Purissima Canyon are generally not as lush or diverse in their understory. Southern sites to visit include Big Basin State Park in the Santa Cruz Mountains north of Santa Cruz, and Pfeiffer-Big Sur State Park, about 30 miles south of the Monterey Peninsula on Highway 1. Redwood forests end altogether in southern Monterey County.

CHAPTER 3

COASTAL SAGE SCRUB: SOUTHERN CALIFORNIA'S "SOFT" CHAPARRAL

Coastal sage scrub, or "soft" chaparral, is the step-sister to the better-known "hard" chaparral (see chapter 10), but in overall structure it is really closer to north coastal scrub. Slathering bluffs and hills close to the sea in a gray-green blanket from Monterey County south into Baja California, coastal sage scrub wanders inland where fogs linger or temperatures remain mild through the year. Coastal sage scrub may also rapidly, if temporarily, invade adjacent hard chaparral or oak woodland after fire. Despite the dominance of coastal sage scrub over a large domain, many ecologists consider it one of California's most threatened plant communities because of the phenomenal urban sprawl in coastal southern California.

The first impression of this assemblage of small, loosely knit shrubs may be disappointing in summer or fall, when many of the shrubs have lost most of their leaves to drought and are semidormant. With the resurgence of winter rains, however, the shrubs are magically rejuvenated with vigorous sets of fresh leaves. Happily, most of these shrubs can be maintained in full health in the garden by means of occasional summer water and excellent drainage.

Coastal sage shrubs display a variety of intriguing leaf designs. Soon after new leaves appear, colorful, bee-pollinated flowers may also make their debut, lasting well into summer in years with cool, damp springs. These flowers light up garden spaces.

Coastal sage is noted for its abundance of fast-growing, opportunistic shrubs and succulents. Cacti and dudleyas grow well in this arid habitat too, where the average rainfall is often less than fifteen inches, sometimes as little as ten. Coastal sage is noted also for the strong perfumes of its leaves. In fact, the community's name comes from two different aromatic sources: species of the sagebrush genus, *Artemisia*, in the daisy

LEFT: Small floriferous shrubs include southern bush monkeyflower (*Mimulus longiflorus,* orange) and golden yarrow (*Eriophyllum confertiflorum,* yellow). • ABOVE: The flowers of bladderpod (*Isomeris arborea*).

family; and species of sage, or *Salvia,* belonging to the mint family. Besides these, other members of the daisy, mint, sumac, and rue families add their own special aromas. Because of the plants' oil-impregnated leaves, coastal sage scrub easily catches fire; it is best to plant the shrubs away from buildings, with a watered zone between.

Creating a Garden

Coastal sage scrub is a perfect choice for gardens along southern California's coast, allowing you to recreate lost habitat where most homeowners have not been so far-sighted. You might, for example, combine two or three kinds of shrubs in large containers to create beautiful miniportraits of this special habitat. Elements from the coastal

Mediterranean-themed garden with coastal sage plants.

sage scrub community are also ideal in an herb garden, where strong and pleasant fragrances help create a special ambiance. "Hard" chaparral shrubs can be mixed in to create a woody border, and various bunchgrasses, perennials, and bulbs can provide nice accents as well.

Design Notes

Creating this design gave me the opportunity to imagine how the city of Los Angeles might be garden-retrofitted with plants from coastal sage scrub, a natural plant community that thrived in that city prior to urbanization. These plants are superimposed on the existing urban grid of concrete, steel, glass, and stucco.

One drawback to living in an urban high-rise is the lack of trees nearby and the fact that views of nature are far off. To remedy this, I have designed tree columns for

tall buildings—which, in seismically active areas, may actually contribute over time to the building's stability as well. The hollow columns of steel, with fiberglass linings, are filled with soil; they are also left open to the ground, so that roots can penetrate the earth below. There is an opening every fifteen vertical feet, in which large shrubs such as lemonade berry and laurel sumac are planted as seedlings. (Other species used in this particular design are summer holly [*Comarostaphylis diversifolia*], island

CLOCKWISE FROM TOP LEFT: Coastal sage garden in a rustic setting. • A cottage garden with coastal sage scrub plantings • White sage (center) (*Salvia apiana*) with yarrow (*Achillea millefolium*) and other sages. Photograph by S. Ingram.

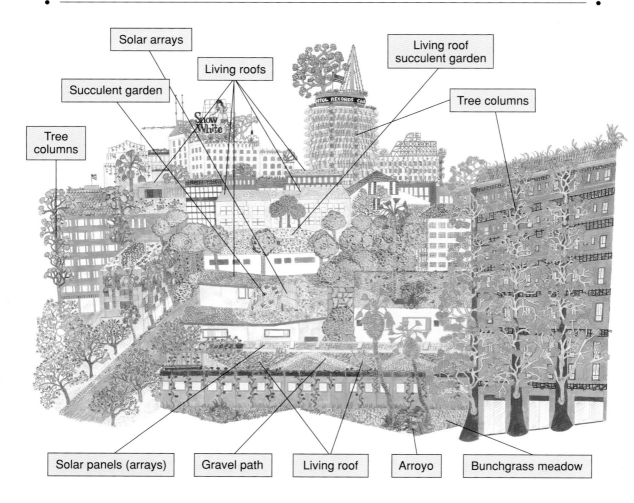

Solar arrays

Living roofs

Succulent garden

Living roof
succulent garden

Tree columns

Tree
columns

Solar panels (arrays)

Gravel path

Living roof

Arroyo

Bunchgrass meadow

manzanita [*Arctostaphylos insularis*], and purple sage.) Each opening is framed in rigid plastic at the top and sides to shield the young plant from overhead roots and soil. A noncorrosive drip pipe with emitters placed every three to four feet runs the length of the column from the top to the ground, assuring that each plant will develop healthy roots. Within five years, the shrubs will have grown six to eight feet. Although the shrubs we have chosen are tall, reaching as much as twelve feet in height, their growth can be controlled by the amount of water received, which will limit root development and hence overall size.

Any size trees can be established from seedlings on any high-rise building. The circumference of the column will determine the ultimate height of the tree. As with shrubs, the manner in which the water is supplied to the root zone will affect overall tree size.

Green roof technology allows the establishment of gardens on any roof. The main

South coastal scrub concept garden.

requirement is that a waterproof membrane and a moisture and root barrier be installed prior to planting. Green roofs help cool the insides of buildings, lower the temperatures of cities, and reduce urban runoff. They also can last twice as long as conventional roofs because the roofing materials themselves are not subjected to sun, wind, and water. Depending on the system and the load-bearing capacities of the roof, the planting area will likely be at least 0.8–1.2 inches deep. At such shallow depths, succulents such as sedum are appropriate. Intermediate depths, of two to six inches, will support a wider range of succulents and grasses—such as dudleya, the sedge *Carex pansa*, giant rye grass 'Canyon Prince', and bunchgrasses like purple three-awn, purple needlegrass, and nodding needlegrass—while eight-inch depths will accommodate many drought-tolerant native perennials, including blue foothill penstemon (*Penstemon heterophyllus*), lilac verbena, coast aster (*Aster chilensis*), bush monkeyflower, purple sage, white sage, and blue-witch. Trees such as lemonade berry and laurel sumac require a planting depth of 32–52 inches.

For meadow roof gardens using native grasses, plant individual plugs at twelve-inch intervals. Mulch the area to a depth of four to six inches with organic materials, preferably ProChip, made from used wooden pallets and other recycled garden waste. Irrigate with laser tubing. Meadows can also be irrigated with overhead spray, but this requires more water and maintenance.

As for the plants, once established they will require some summer water, but because they have evolved through natural selection to thrive in southern California, their needs will be significantly less than those of subtropical species. Water can be supplied with durable drip tubing that is buried under a thick layer of light-colored gravel, which doubles as mulch.

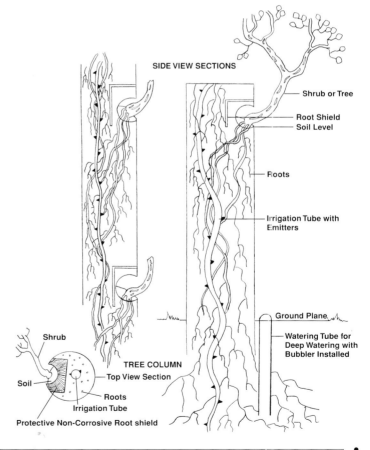

SIDE VIEW SECTIONS

Shrub or Tree

Root Shield
Soil Level

Roots

Irrigation Tube with Emitters

Ground Plane

Watering Tube for Deep Watering with Bubbler Installed

TREE COLUMN
Top View Section

Shrub

Soil

Roots
Irrigation Tube

Protective Non-Corrosive Root shield

Tree column vertical cut away.

Scope of Work

Tree Columns

- ◆ Create tree columns for high-rise structures during the original construction, or retrofit existing structures. All construction materials should have a useful life equal to that of the building itself.
- ◆ Install the irrigation system and the soil from the ground up as the column is constructed.
- ◆ Install seedlings as the column is constructed.

Green Roof

- ◆ Install waterproof membrane, drainage, and irrigation system.
- ◆ Install soil.
- ◆ Install plants.
- ◆ Connect each plant to drip lines or laser tubing.
- ◆ Install mulch.

Concept garden key.

Plant List

Symbol	Botanical Name	Common Name
A	*Agave deserti*	desert agave
B	*A. shawii*	coast agave
C	*Arctostaphylos glandulosa* ssp. *crassifolia*	Del Mar manzanita
D	*A. insularis*	island manzanita
E	*Calystegia* 'Anacapa Pink'	southern morning glory
F	*Ceanothus papillosus*	wart-leaf ceanothus
g	*Comarostaphylis diversifolia*	summer holly
H	*Coreopsis maritima*	sea dahlia
I	*Dudleya lanceolata*	lance-leaf dudleya
J	*D. pulverulenta*	chalk dudleya
K	*Eriogonum fasciculatum*	California buckwheat
L	*Eriophyllum confertiflorum*	golden yarrow
M	*Encelia farinosa*	coast brittlebush
N	*Galvezia speciosa*	island snapdragon
O	*Isomeris arborea*	bladderpod
P	*Leymus condensatus* 'Canyon Prince'	giant rye grass
Q	*Malacothamnus fasciculatus*	southern bush mallow
R	*Malosma laurina*	laurel sumac
S	*Mimulus longiflorus*	southern bush monkeyflower
T	*Nassella cernua*	nodding needlegrass
not shown	*N. pulchra*	purple needlegrass
U	*Ornithostaphylos oppositifolia*	Baja birdbush
f	*Quercus agrifolia*	coast live oak
V	*Rhus integrifolia*	lemonade berry
W	*R. ovata*	sugar bush
X	*Salvia apiana*	white sage
Y	*Salvia clevelandii*	Cleveland sage
Z	*S. leucophylla* 'Pt. Sal'	purple sage
a	*Sedum spathulifolium*	common stonecrop
b	*Solanum xanti*	blue-witch
c	*Verbena lilacina*	lilac verbena
d	*Vitis girdiana* or *V. californica* 'Roger's Red'	desert grape or California grape
e	*Hesperoyucca whipplei*	chaparral yucca

Plants to Use

Large Shrubs

Lemonade berry (*Rhus integrifolia*).

Dense, multitrunked evergreen shrub or small tree to 20 feet high. Rigid branches carry

tough, nearly round, dark green leaves lined with a few coarse teeth. Dense panicles of small, rose to pale pink blossoms appear in late winter, later in the north. Flattened drupes covered with velvety, sticky red hairs decorate shrubs in midspring; steeping these fruits in water creates a lemonadelike drink. Can be pruned into a hedge or espaliered. Propagate from hardwood cuttings or scarified seed.

Garden Design Note: Can't take as much inland heat as sugar bush (*Rhus ovata*).

Laurel sumac (*Malosma laurina*).

Spreading evergreen shrub or small tree to 20 feet tall, often as broad as tall. Flexible red twigs carry fragrant, glossy, bright green, broadly ovate leaves that are folded upward. Frothy panicles of pinkish buds open to tiny white flowers in early summer, followed by greenish fruits that turn brown by summer's end. Laurel sumac grows farther inland and at higher elevations than lemonade berry. Propagate from hardwood cuttings or scarified seed. Laurel sumac can be shaped into a small tree or allowed to grow as a multitrunked shrub.

Garden Design Note: Makes a fast-growing screen, especially in organically rich soils.

Southern bush mallow (*Malacothamnus fasciculatus*).

Vigorous, fast-growing, sometimes aggressive multibranched shrub to 10 feet high with many stiffly upright branches. Variable rounded leaves, sometimes grayish, sometimes green. Long succession of cup-shaped, hollyhocklike, pink-purple flowers in slender spikes in late spring and early

TOP: Sugar bush (*Rhus ovata*) is similar to lemonade berry but takes more heat and winter cold. • BOTTOM: Flowers of southern bush mallow (*Malacothamnus fasciculatus*).

summer; often reblooms in late summer. Propagate from semihard-wood cuttings, divisions, or seed. Bush mallows often become leggy and can be cut back to reshape; be sure to use sharp pruning shears, as bark tends to tear. Garden Design Note: Overall appearance of pale gray-green contrasts nicely with coffeeberry, ceanothus or manzanita as background plantings. The flowers are plentiful and bloom over a long period.

Smaller Shrubs

California brittlebush (*Encelia californica*).

Low, sometimes semiprostrate, semievergreen shrub to three feet tall with broadly ovate, dark green, scented leaves, raspy to the touch. Large yellow daisies with a dark purple center are borne singly

above foliage. Blooms from spring into early summer. Propagate from semihardwood cuttings or seed. Garden Design Note: Extreme dieback after flowering if water is withheld.

California sagebrush (*Artemisia californica*).

Twiggy, semievergreen shrub with many weak, upright branches, reaching six feet high. A fine prostrate form, 'Canyon Gray', comes from the Channel Islands. Grayish green leaves are finely divided into linear segments and are heavily scented. Rather inconspicuous spikes of tiny, pale yellowish flowers emerge in fall. California sagebrush is prominent in both north coastal and coastal sage scrubs. Propa-

TOP: Coastal sage scrub garden on steep slope with manzanita handrail. • BOTTOM: California sagebrush (*Artemisia californica*) has a feathery texture.

gate from hardwood cuttings or seed. To improve shape, California sagebrush can be cut back by half in the fall. Removing old flowering stalks improves appearance.

Garden Design Note: Weave together California sagebrush and sticky monkeyflower on a hillside for a dramatic effect.

White sage (*Salvia apiana*).

White sage is among our most dramatic salvias, with low, dense branches seldom more than three

TOP: Chaparral yucca (*Hesperoyucca whipplei*) amongst white sage (*Salvia apiana*). • BOTTOM: A Mediterranean garden featuring chaparral plants.

feet high and large, white, oval, strongly scented leaves. Wandlike flowering stalks to six feet high appear in late spring and summer with many close clusters of showy white or palest purple, two-lipped flowers attractive to bees. Propagate from cuttings or seed. White sage needs a minimum of maintenance, chiefly the removal of old flowering stalks.

Garden Design Note: Give white sage lots of room. It doesn't behave well in small spaces, but it can be spectacular on a hillside, especially in evening light. The hybrid 'Vicki Romo' is smaller.

Purple sage (*Salvia leucophylla*).

Much-branched shrub to six feet high with narrowly elliptical, gray-green, quilted leaves of less pervasive smell than white sage. Early-summer whorls of pale lavender to purple, two-lipped flowers are a beautiful foil to the foliage. Another great bee plant. The best forms are tight and dense, such as 'Pt. Sal', or hybrids with the low-growing Sonoma sage (*S. sonomensis*). Prune heavily to shape. Propagate from cuttings.

Garden Design Note: Sages are the workhorses of habitat gardens, attracting numerous pollinators and bird species.

Black sage (*Salvia mellifera*).

Rather low, densely branched, broadly rounded, semievergreen shrub to about four feet high, though some forms are nearly prostrate. Strongly scented, narrowly elliptical, dark green leaves, paler underneath. Whorls of palest purple to nearly white, two-lipped flowers in short spikes in spring. A bee favorite and excellent source of honey. Defoliates in hot summers. Propagate from semihardwood cuttings or seed. Black sage can be sheared back to promote new growth, improve density, and reduce legginess.

Garden Design Note: 'Tera Seca' is low (one to two feet) and spreading (six feet). Contrasts nicely with 'Bee's Bliss'.

Bladderpod (*Isomeris arborea*).

Broad, fast-growing, multibranched, semievergreen shrub to about eight feet tall. Bluish green, trifoliate, ill-scented leaves and tight racemes of golden, four-petaled flowers much of the year.

Flowers and seed pods of bladderpod (*Isomeris arborea*).

93

Flowers are followed by conspicuously inflated, green, balloon-shaped seed pods. Garden Design Note: Not as available as it should be. It is a tireless performer. I have big yellow blossoms on my bladderpod nearly every month of the year.

Blue bush or silver-leaf lupine (*Lupinus albifrons*).
Fast-growing, semievergreen shrub to three or four feet tall with gray to silvery, palmately compound leaves. Spikes of fragrant blue or purple (occasionally white), sweet pea-like flowers in midspring. Attractive to many pollinators. Propagate from fresh seed. Plants tend to be short-lived in gardens; expect three to four years. Give excellent drainage to prevent root rot.

Prickly phlox (*Leptodactylon californicum*).
Low, evergreen, multibranched shrub to three feet high with sharply prickly, fingerlike leaf lobes and clusters of showy, bright pink, phloxlike flowers in late winter and spring. Propagate from seed. Give full sun. Grows slowly and needs perfect drainage.
Garden Design Note: Try this in a container first, to get the drainage right.

Southern bush monkeyflower (*Mimulus longiflorus* [= *M. aurantiacus* in *The Jepson Manual*]).
Fast-growing, spreading shrublet to two or three feet tall with narrow, sticky leaves and a long succession of showy, two-lipped, pale yellow or orange flowers attractive to bees. Closely related to the northern bush monkeyflower, *M. aurantiacus*. Defoliates in hot dry summers. With water, blooms from spring into summer. Propagate from tip cuttings or seed. Deadheading may encourage more flowering stalks; tip pruning will improve bushiness. Also try red-flowered *M. puniceus*.
Garden Design Note: Excellent garden companion to *Lupinus albifrons*. Plant them together on a hillside or a constructed garden mound.

LEFT: Southern bush monkeyflower (*Mimulus longiflorus*) in peak bloom. • RIGHT: Bush monkeyflower (*Mimulus* 'Paprika') and white sage (*Salvia apiana*).

Blue-witch (*Solanum xanti*).

Densely branched, semiwoody shrub to three feet high with felted, sticky, ovate leaves and umbel-like clusters of fragrant blue-purple flowers. Flowers have a flat starlike shape with a center of yellow stamens. Blooms from spring to summer. Defoliates in hot dry summers. Propagate from seed or semihardwood cuttings. Can be pruned back to encourage bushiness.

Garden Design Note: Mass blue-witch with California bush poppy (*Dendromecon rigida*) as a focal point, or with narrow-leaved goldenbush (*Ericameria linearifolia*). Don't plant in a children's garden; it is a member of the nightshade family and therefore poisonous.

Berry-rue (*Cneoridium dumosum*).

Dense shrublet to four feet high, though it may be pruned to keep tight and low. Shiny, evergreen leaves are imbued with a bitter rue odor. Small clusters of pink-budded, white, citruslike flowers appear in winter and early spring, followed by rose-red, berry-like fruits. Propagate from cuttings or seed.

Garden Design Note: Not readily available. Plant under desert willow (*Chilopsis linearis*) as a tall ground cover in hot climates.

California buckwheat (*Eriogonum fasciculatum*).

Dense shrub to three feet tall with clusters of narrow, dark green leaves. Flat-topped to rounded clusters of pink-budded, white flowers emerge in late spring and summer, becoming rust colored. Propagate from hardwood cuttings or seed. Old flower heads make pleasing dried arrangements; removing them encourages more flowering and improves the appearance of plants.

Garden Design Note: Excellent fall border plant. Annual pruning keeps this plant from getting out of control. Excellent mass planted on a hillside; great for erosion control.

Perennials

Sea dahlia (*Coreopsis maritima*).

Woody-based perennial to two feet high with tufts of finely divided, fleshy, fernlike leaves. Long series of large, yellow daisies from winter through spring. Tender in winter and vulnerable to chewing insects; protect against the predations of snails and slugs. Propagate from seed. Plants tend to be short-lived but grow rapidly from seed to replace previous specimens.

Flowers of blue-witch (*Solanum xanti*).

Garden Design Note: A bright spot of color, with flowers that are large and plentiful.

Lilac verbena (*Verbena lilacina*).
Woody-based, rounded perennial that creates tufts two or three feet across and high with small, deeply pinnately lobed, evergreen leaves and headlike spikes of sweetly

scented, lavender flowers. Flowers almost continuously in mild winters. Winter tender but resprouts from roots if cold is not prolonged. Propagate from cuttings. Lilac verbena suffers from excessive pruning; never cut into old wood. Tip pruning improves bushiness.

Garden Design Note: This native of Isla Cedros off Baja California is one of my favorite flowering perennials. It blooms for months on end, thrives in the hot sun, is fast-growing, and butterflies love it. 'De la Vina' is a good selection.

Goldenstars (*Bloomeria crocea*).
Late spring-blooming bulbs with foot-tall umbels of bright yellow, starlike flowers that open after leaves dry. Mass for effect. Propagate from seed or cormlets.
Garden Design Note: Difficult to find but worth the effort. Show-stealing flowers stand out against the soil with no leaves to compete.

California peony (*Paeonia californica*).
Late summer/fall-dormant perennial to three feet high with peonylike leaves and small clusters of nodding, dark red flowers in late winter and early spring. Needs perfect drainage and high shade. Withhold summer water. Propagate from seed.

Garden Design Note: Another rare plant in the nursery trade. Unique foliage with subtle flowers unlike garden-variety peonies.

Southern morning glory (*Calystegia macrostegia*).
Fast-growing, winter-dormant vine with arrowhead-shaped leaves and large, funnel-shaped white or pink flowers that open from spring through sum-

TOP: Lilac verbena (*Verbena lilacina*) in full flower. • BOTTOM: Flowering stems of southern morning glory (*Calystegia macrostegia*).

mer. Propagate from soaked or stratified seed. 'Anacapa Pink' is a superior cultivar. Morning glory can be cut back to the woody stems in fall to encourage new growth in spring and avoid interference with the old growth.

Garden Design Note: Very fast grower but may be short-lived in heavy clay. Will cover an unattractive fence in a season or two.

Bush savory (*Satureja chandleri*).

Small, semievergreen, woody-based perennial to 18 inches high with highly fragrant, rounded, dull green leaves and clusters of small white, two-lipped flowers in late spring and summer. Needs excellent drainage and is best with high shade. Propagate from semihardwood cuttings.

Mixed coastal sage garden with lilac verbena (*Verbena lilacina*) in the background. Photograph by S. Ingram.

Garden Design Note: Available in specialty nurseries in southern California. Use as a specimen in a perennial garden.

Succulents

Shaw's agave (*Agave shawii*).
Bold rosettes three feet across of deep green, broadly ovate leaves are bordered by wickedly recurved spines. Fat flower stalks carry panicles of tubular, yellow flowers to ten feet high. Flowers appear only once after eight or more years of growth. Propagate from offsets or seed. Remove dead flower stalks and leaf rosettes. Old clumps should be divided, as they become dense.
Garden Design Note: I've seen six- to ten-foot flower stalks decorated as Christmas trees in Baja roadside bars and restaurants. Native Americans baked the young flower buds in large pit ovens each spring.

Britton's dudleya (*Dudleya brittonii*).
Dramatic rosettes two feet across of grayish white, lance-ovate leaves, with cymelike spikes of pale yellow flowers in late spring or early summer. Needs excellent drainage. Propagate from seed. Native to northern coastal Baja California.
Garden Design Note: The most dramatic of the native dudleyas. Use as an accent in a coastal rock garden. Plant in loose gravel or try in a container first.

Grasses

Purple three-awn (*Aristida purpurea*).
Small bunchgrass to two feet high with narrow leaves and brushy, spikelike clusters of

purple-tinted flowers through much of the year. Decorative flower heads. Propagate from seed or divisions.

Giant rye grass (*Leymus condensatus*).
Bold clumps of leafy stems to eight feet tall, topped by fat spikes of flowers in summer. A commanding accent plant for full sun or light shade. The best form is the cultivar 'Canyon Prince', which is only four feet tall and has bluish green leaves. Propagate from divisions or seed. Rye grass may need to be thinned periodically; be sure to wear gloves, for the leaves are abrasive and easily cut the skin.
Garden Design Note: This is one of my favorite grasses for bold architectural gardens. Mass

The bold flower spikes of giant rye grass (*Leymus condensatus* 'Canyon Prince').

Craftsman-style garden with *Calystegia* 'Anacapa Pink' and island bush poppy (*Dendromecon harfordii*). Photograph by S. Ingram.

'Canyon Prince' next to *Festuca idahoensis* 'Siskiyou Blue' or leafy reed grass (*Calamagrostis foliosa*).

Additional large shrubs to try:

Baja birdbush (*Ornithostaphylis oppositifolia*), chaparral yucca (*Hesperoyucca whipplei*).

Additional small shrubs to try:

Tree coreopsis (*Coreopsis gigantea*), golden yarrow (*Eriophyllum confertiflorum*), California brickel bush (*Brickellia californica*), climbing penstemon (*Keckiella cordifolia*), island snapdragon (*Galvezia speciosa*), showy penstemon (*Penstemon spectabilis*), Catalina lace (*Eriophyllum nevinii*).

Additional perennials to try:

Wishbone bush (*Mirabilis californica*), southern pink (*Silene laciniata*), purple needlegrass (*Nassella pulchra*), nodding needlegrass (*N. cernua*), lance-leaf dudleya (*Dudleya lanceolata*), blue dicks (*Dichelostemma capitatum*).

Giant rye grass (*Leymus condensatus* 'Canyon Prince') in a coastal planting. Photograph by A. Owens.

Places to Visit

To see coastal sage scrub in its native habitat, visit the coastal hills by Pt. Sal south of Vandenberg Air Force Base in northern Santa Barbara County, the steep frontal slopes of the Santa Ynez Mountains behind Santa Barbara, or the mountains behind Ojai and Santa Paula in Ventura County. These places feature beautiful mosaics of coastal sage scrub, chaparral, and oak and riparian woodlands. Other sites include:

Montaña de Oro State Park, San Luis Obispo County. South of Morro Bay, take Hwy 1 to Montaña de Oro State Park. The section overlooking the ocean features a combination of coastal chaparral and coastal sage scrub, with Morro manzanita (*Arctostaphylos morroensis*), mock-heather (*Ericameria ericoides*), Blochman's senecio (*Senecio blochmanae*), California sagebrush (*Artemisia californica*), black sage (*Salvia mellifera*), rushrose (*Helianthemum scoparium*), little-leaf buckwheat (*Eriogonum parvifolium*), beach lupine (*Lupinus chamissonis*), deerbroom lotus (*Lotus scoparius*), golden yarrow (*Eriophyllum confertiflorum*), and more.

Camino Cielo, Santa Ynez Mountains, Santa Barbara County. From Hwy 152 north of Santa Barbara, take Painted Cave Road to the top, then turn right onto Camino Cielo and continue on. The ridge-top is a mixture of coastal sage scrub and chaparral, the scrub often replacing chaparral after burns or on substrates less amenable to hard-chaparral species.

Prominent plants include deerbroom lotus, canyon sunflower (*Venegasia carpesioides*), woolly blue-curls (*Trichostema lanatum*), sawtooth goldenbush (*Hazardia squarrosa*), prickly phlox (*Leptodactylon californicum*), black sage, white sage (*Salvia apiana*), laurel sumac (*Malosma laurina*), sugar bush (*Rhus ovata*), chaparral yucca (*Hesperoyucca whipplei*), golden yarrow, and woolly yerba santa (*Eriodictyon tomentosum*).

Torrey Pines State Park, San Diego County. The park lies on the coast south of Del Mar, north of San Diego, and west of Hwy 5. The vegetation is a combination of maritime chaparral, coastal sage scrub, and Torrey pine forest. Prominent shrubs include bladderpod (*Isomeris arborea*), Nevin's barberry (*Berberis nevinii*), California sagebrush, sea dahlia (*Coreopsis maritima*), berry rue (*Cneoridium dumosum*), lemonade berry (*Rhus integrifolia*), mission manzanita (*Xylococcus bicolor*), wartleaf ceanothus (*Ceanothus verrucosus*), black sage, and Shaw's agave (*Agave shawii*).

Coastal vegetation, Montaña de Oro State Park, San Luis Obispo Co.

CHAPTER 4

CHANNEL ISLANDS GARDEN: A PARADE OF UNIQUE PLANTS

Floating dreamlike off the southern California coast, the eight Channel Islands run the gamut from rocky islets to, in effect, minicontinents. Surrounded by a moderating ocean, they host a wide variety of plants found nowhere else, ranging from ancient oaks and pines to unique ironwoods, striking chaparral shrubs, and giant herbaceous perennials. Although feral animals have done untold damage to these fragile floras, removal efforts have led to a surprising rebound of the native vegetation.

Though Santa Catalina Island is famed for its easy access, glass bottom boat tours, and welcoming accommodations, it's Santa Cruz Island, 20 miles off the Santa Barbara coast, that best exemplifies the Channel Islands' biological uniqueness, with a broad array of handsome endemics. Santa Cruz Island has its own central valley and two mountain ranges, with Picacho Diablo rising to 2,300 feet in elevation.

The plants and animals that have made the trip across the water have evolved and adapted in their own ways, resulting in unique communities. Newly evolved species join company with relicts that long since died out on the mainland, but there are also many species that occur both on the mainland and in the islands. Juxtapositions of pine forests, ironwood groves, coastal sage scrub, and chaparral create complex mosaics that reflect the mild year-round conditions of a classic Mediterranean climate region.

LEFT: Pelican Bay on the north side of Santa Cruz Island with coast live oak (*Quercus agrifolia*) and tree coreopsis (*Coreopsis gigantea*). • THIS PAGE, TOP: Tree coreopsis on rugged cliffs. • BOTTOM: Flowering mound of prostrate chamise (*Adenostoma fasciculatum* var. *prostratum*).

Creating a Garden

If you live in the mild climates along our coast—especially in southern California—a garden design incorporating a mix of island elements is both appropriate and beautiful, evoking a place of remote and singular flavor. By mixing the components of a rock outcrop or a stony wash with a copse of trees, say, you can grow a diversity of beautiful island plants from different environmental niches. Pay attention, however, to which plants require shade and which full sun.

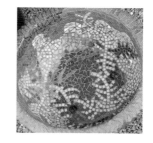

As with most designs, good drainage is essential, but it is

TOP: Craftsman style garden with island bush poppy (*Dendromecon harfordii*) providing structure and color. BOTTOM: Artist Christina Yaconelli created this mosaic sink featuring Catalina lace (*Eriophyllum nevinii*).

particularly critical for plants that favor rock scree and dry washes. Examples include the Santa Cruz Island buckwheat (*Eriogonum arborescens*) and scale-broom (*Lepido-spartum squamatum*). Many island plants are winter cold-tender, although many resprout from roots when frosts don't last too long. The plants listed below are presented according to habitat. Please note that not all of these plants are restricted to the islands; unique plants are noted as *endemic*.

TOP LEFT: Pink alum root (*Heuchera* sp.) and California fescue (*Festuca californica*) in a rustic garden setting. • TOP RIGHT: Flagstone pathway with mounded plantings of Catalina perfume (*Ribes viburnifolium*) and glorymat (*Ceanothus gloriosus*). • BOTTOM: Pink buckwheat (*Eriogonum grande* var. *rubescens*) in a Channel Island garden.

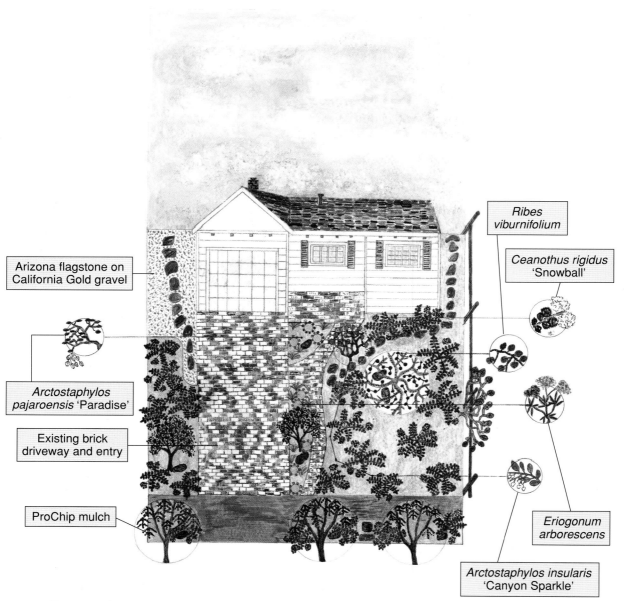

Arizona flagstone on California Gold gravel

Ribes viburnifolium

Ceanothus rigidus 'Snowball'

Arctostaphylos pajaroensis 'Paradise'

Existing brick driveway and entry

Eriogonum arborescens

ProChip mulch

Arctostaphylos insularis 'Canyon Sparkle'

Design Notes

A significant percentage of the gardens we create are for single-story tract homes in older neighborhoods, such as this two-bedroom, wood-siding house in San Jose. The home was owned by a retired couple who enjoyed spending time with their grandchildren and were tired of pushing a lawn mower.

Rendering of featured garden.

The first time I met our new clients, I had recently returned from a botanizing trip to Santa Cruz Island. Plants indigenous to that island offered a natural palette for their site, which received full southwestern sun in the front. One of my favorite California natives, *Eriogonum arborescens* (Santa Cruz Island buckwheat), was in full bloom on my trip, and I was anxious to introduce it into this small garden in the company of its ecological neighbors.

To the recently installed brick driveway and walk, we added Arizona flagstone pathways and Arizona limestone accent boulders. We removed the turf and created undulating berms 12 to 18 inches high to accommodate the new chaparral and coastal bluff plantings.

I settled on several species and hybrids of *Arctostaphylos* and *Ceanothus*. In making my selections, I utilized the helpful notes in the Tree of Life Nursery catalog. Because it is often difficult to locate particular species of ceanothus ground cover, in the plant list below I have offered several suggestions. In selecting one, it is important to know the mature size of the plant and space accordingly for optimum growth. (Most people overplant native gardens, which creates needless garden waste and gives natives a bad reputation. When a native garden is first established, there should be large empty spaces between the young plants. They grow very fast. The pleasure is in observing the garden become a habitat.) A thick covering of mulch will keep weeds in check, and drip irrigation will help as well.

This garden is full of trees and shrubs that are in flower most of the year. Colors are predominantly blue, white, yellow, rust, dusty pinks, and lavender.

Tufa planters are filled with a collection of showy perennials—various dudleyas, buckwheats, and island snapdragon (*Galvezia speciosa*): miniature rock gardens that can be displayed in the garden or at the front entry. Brushing up against a fragrant pitcher sage will fill you with the heady aroma of a wild California island. Using these delightful plants to replace a sterile, water-guzzling lawn is a satisfying experience.

TOP TO BOTTOM: *Ribes viburnifolium* • Catalina lace (*Eriophyllum nevinii*) • Santa Cruz Island buckwheat (*Eriogonum arborescens*) • Tree ceanothus (*Ceanothus arboreus*). Illustrations by A. Yankellow.

Scope of Work

- ◆ Remove turf and unwanted plants.
- ◆ Prune and shape existing trees that are to remain.
- ◆ Grade to drain rainwater away from path yet retain it on site.
- ◆ Build a series of berms to accommodate bluff and chaparral plantings.
- ◆ Build flagstone paths on a bed of decomposed granite.
- ◆ Install Arizona limestone boulders.
- ◆ Install drip irrigation.
- ◆ Install curbside and driveway plantings.
- ◆ Install tufa planters, running drip lines to each planter.
- ◆ Mulch with three inches of brown ProChip.

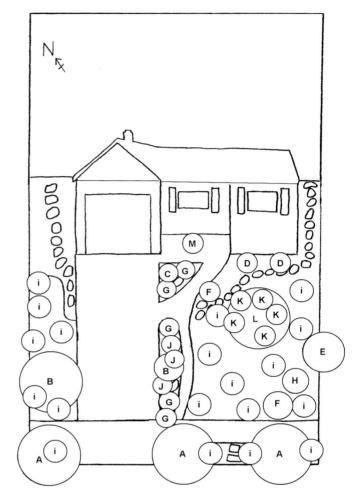

Plant List

Symbol	Botanical Name	Common Name
A	*Lyonothamnus floribundus* ssp. *aspleniifolius*	Santa Cruz Island ironwood
B	*Ceanothus arboreus*	tree ceanothus
C	*Arctostaphylos pajaroensis* 'Paradise'	Pajaro manzanita
D	*Ceanothus rigidus* 'Snowball'	snowball ceanothus
E	*Rhus ovata*	sugar bush
F	*Arctostaphylos insularis* 'Canyon Sparkle'	canyon sparkle manzanita
G	*Eriogonum arborescens*	Santa Cruz Island buckwheat
H	*Arctostaphylos catalinae* or *A. edmundsii* 'Carmel Sur'	Catalina manzanita

Featured garden key.

i	*Ceanothus maritimus*, *C. hearstiorum*, or *C.* 'Anchor Bay'	maritime ceanothus
J	*Festuca californica*	California fescue
K	*Ribes viburnifolium*	evergreen currant
L	*Prunus ilicifolia* ssp. *lyonii*	island cherry
M	Tufa (stone sink) container garden plants, to be selected from:	
	Lepechinia fragrans	fragrant pitcher sage
	Keckiella cordifolia	climbing penstemon
	Eriogonum grande var. *rubescens*	rose buckwheat
	Dudleya candelabrum	candelabra dudleya
	Eriophyllum nevinii	Catalina lace
	Galvezia speciosa	island snapdragon

Plants to Use

Woodland Plants

Island oak (*Quercus tomentella*), ENDEMIC.
Tall, shapely tree to 50 feet high, sometimes with spreading crown and often multi-trunked. Large, deep green, lance-shaped leaves with serrated margins, woolly beneath; catkins of male flowers in early spring; ripe acorns in fall. Propagate from fresh acorns. Island oak needs little maintenance. Garden Design Note: Greatly underutilized urban tree. Excellent street tree; also ideal for groves.

Santa Cruz Island or fern-leaf ironwood (*Lyonothamnus floribundus* ssp. *aspleniifolius*), ENDEMIC.
Tree to 40 feet high that vigorously suckers to form colonies. Beautiful, ribbonlike, red-brown bark and dark green, fernlike, evergreen leaves. Flat-topped clusters of small white roselike flowers appear in late spring. Brown seed pods stay on trees for several years. Propagate from suckers or stratified

Santa Cruz Island ironwood (*Lyonothamnus floribundus* ssp. *aspleniifolius*). Photograph by S. Ingram.

seed. Because of the vigorous suckers, care should be exercised to remove all unwanted growth. The thirsty, shallow roots make it difficult to underplant without extra water. Garden Design Note: This distinctive tree lends great beauty to the landscape. Excellent in a grove, where you can leave the shaggy bark strips and fern leaf litter as natural mulch.

Island snapdragon (*Galvezia speciosa*), ENDEMIC.
Sprawling evergreen shrub, which can climb to eight feet. Small oval leaves. Racemes of bright red snapdragon flowers bloom much of the year. Requires well-drained soil, light shade, no or little water; may be cold-tender. Propagate from cuttings. Island snapdragon can be heavily pruned to keep it bushy; stems, which tend to get leggy, are amenable to being trained as an espalier.

Garden Design Note: This plant can provide late-blooming, bright red color in the fall garden. 'Firecracker' is especially vibrant and compact.

Climbing penstemon (*Keckiella cordifolia*).
Woody climber to 15 feet high. Small, ovate, toothed leaves and long racemes of wide two-lipped, scarlet (sometimes orange or yellow) flowers in summer. Attractive to hummingbirds. Propagate from semihardwood cuttings or seed. Seeds may volunteer in gardens.
Garden Design Note: Plant this climber under shrubs and allow into the crown to blossom. I like to see it ramble up a nearby toyon or 'Bonita Linda' coffeeberry.

Evergreen currant or Catalina perfume (*Ribes viburnifolium*), ENDEMIC.
Clambering shrub with branches that swoop up, then arch over and down. Glossy, round, fragrant, evergreen leaves, with clusters of tiny dark red flowers in spring. Propagate from layered branches or hardwood cuttings; pinning stems to the ground helps to promote new plants or establish a colony. May be used as a loose, woody ground cover.

Island alumroot (*Heuchera maxima*), ENDEMIC.
Large version of coralbells, to two feet tall, with bold rosettes of nearly round, scalloped leaves and panicles of creamy, bell-shaped flowers. Blooms spring and early summer. Older plants may develop conspicuous woody stalks below the tufts of leaves; cut them off and use to start new plants.

Island snapdragon (*Galvezia speciosa*) planted as a free-form hedge.

Garden Design Note: Effective massed, especially with other selections of heuchera. Excellent under oaks with native bunchgrasses.

Humboldt lily (*Lilium humboldtii* var. *ocellatum*).
Bold bulb to eight feet tall (may need support) with whorls of shiny leaves and 20 or more nodding, tiger lily-like, orange flowers with dark spots. Stunning, but vulnerable to deer. Propagate from bulb scale cuttings or seed. Withhold water after blossoms have faded.
Garden Design Note: Plant under oaks with California fescue and irises to mask fading foliage after flowering.

Canyon sunflower (*Venegasia carpesioides*).
Winter-dormant perennial to six feet tall with pairs of bright green, broadly triangular, toothed leaves and close clusters of bright yellow daisies in spring and early summer. Propagate from seed or cuttings. Needs occasional summer water to thrive. After the roots are well established, the plants may be cut back to encourage a second round of growth and bloom.
Garden Design Note: This large, underutilized plant, with its very showy flowers, performs well in sun and shade, and can be grown at the edge of oaks. Available at southern California specialty nurseries.

Island bush monkeyflower (*Mimulus flemingii* [= *M. aurantiacus* in *The Jepson Manual*]), ENDEMIC.
Small evergreen shrub to 15 inches high with pairs of narrow, shiny leaves and racemes of red, two-lipped flowers attractive to hummingbirds. Often hybridizes with southern bush monkeyflower (*M. longiflorus*) to produce flowers of beautiful intermediate colors. Propagate from semihardwood cuttings or seed.
Garden Design Note: Rare in the trade. Can substitute other mimulus hybrids.

Chaparral Plants

Tree ceanothus (*Ceanothus arboreus*), ENDEMIC.
Large shrub or small tree to 20 feet tall, usually with multiple trunks. Evergreen leaves to four inches long, bright green above, silvery beneath. Long sprays of fragrant, pale lavender-blue flowers appear in early spring,

Flowers of tree ceanothus (*Ceanothus arboreus*).

followed by dark purple, three-sided capsules. Propagate from hardwood cuttings or stratified seed. Take care to prune out old dead growth to improve air circulation to the crown.
Garden Design Note: Excellent, fast-growing, small flowering tree for the garden. Good for screening. Plant on a mound with boulders. 'Ray Hartman' or 'Trewithen Blue' are readily available hybrids that can be substituted for *C. arboreus*.

Island mountain mahogany

(*Cercocarpus betuloides* var. *blanchiae*), ENDEMIC.
Large, multitrunked shrub to 25 feet high with smooth, gray bark; small, obovate, toothed, nearly evergreen leaves; racemes of creamy, saucer-shaped flowers filled with multiple stamens; and white-plumed, achene-like fruits. Fruits ripen in summer and, when backlit, glow like candles. Propagate from stratified seed or hardwood cuttings. Remove extra suckers to open up the overall structure.
Garden Design Note: Rare in the trade. Can substitute any mountain mahogany. I've trained it as a formal hedge, as well as artistically pruned and backlit it in an Asian garden.

Island manzanita (*Arctostaphylos insularis*), ENDEMIC.
Large shrub or small, multitrunked tree to 20 feet high. Glossy, dark red bark and green leaves. Clusters of white to pale pink, urn-shaped flowers bloom in early spring, followed by mahogany-red berries. Propagate from hardwood cuttings or scarified seed. Old dead twigs should be pruned out to improve air circulation and help reveal the handsome bark.
Garden Design Note: Outstanding specimen plant. 'Canyon Sparkles' is readily avail-

Tree ceanothus (*Ceanothus arboreus*) underplanted with glorymat ceanothus (*C. gloriosus*) in the featured garden.

able and, at four feet tall and six feet wide with a stout trunk, an adaptable size for the urban garden.

Island cherry (*Prunus ilicifolia* var. *lyonii*), ENDEMIC.
Large, multitrunked shrub or small tree to 30 feet tall with dark bark and large, ovate, smooth-margined, glossy green leaves. Narrow racemes of fragrant, white, cherrylike blossoms appear in profusion in mid to late spring, followed by large, red-purple stone fruits in fall. Fruits are sweet but have a thin pulp. Propagate from seed. Be sure to watch for unwanted suckers at the base of the plants and cut them out.
Garden Design Note: Excellent small tree for urban gardens. Grows fast. Creates a screen that can be sheared or allowed to grow naturally.

Summer holly (*Comarostaphylis diversifolia*).
Very large shrub or small tree with broadly spreading crown to 30 feet tall. Strips of brownish bark and tough, sometimes curled, dark green, elliptical leaves lined with finely serrated teeth. Terminal racemes of nodding, urn-shaped, white flowers in late winter/ early spring and often again in fall. Bright red, warty berries emblazon shrubs in summer. Propagate from hardwood cuttings (bottom heat) or stratified seed. Summer-holly can be shaped into a tree by removing suckers and pruning back hard to force new growth when the old growth shows signs of fungal infection.

Garden Design Note: This manzanita-like tall shrub deserves a place in the late-summer border. Plant it with hummingbird fuchsias and buckwheats, then prune artistically to expose its beautiful bark.

Mission manzanita (*Xylococcus bicolor*).
Densely branched evergreen shrub to eight feet tall with tough, tightly curled under, dark green leaves felted white underneath. Salmon-colored bark and short racemes of pale yellow, lanternlike flowers in late winter, followed by dry, nearly black berries. Propagate from hardwood cuttings or stratified and scarified seed. Prune out old dead twigs to improve air circulation to the crown.
Garden Design Note: Sometimes available at specialty nurseries in southern California. Looks like a cross between manzanita and coffeeberry. Must have excellent drainage. Try it in a container first.

Summer holly (*Comarostaphylis diversifolia*). Photograph by S. Ingram.

Island bush poppy (*Dendromecon harfordii*), ENDEMIC.

Multibranched shrub to eight feet tall that suckers, with ovate, bluish green evergreen leaves. A long succession of three-inch-broad yellow poppy flowers, peaking in spring. Propagate from divisions or stratified seed (or give seed fire treatment).

Garden Design Note: Once established, this is one of the most dependable ever-blooming shrubs. Likes coastal influence. Difficult to establish during autumn months. Plant it with ceanothus, lupines, blue-witch, or in a dry meadow with blue-eyed grass and Ithuriel's spear.

Catalina lace (*Eriophyllum nevinii*), ENDEMIC.

Semievergreen shrub to four feet tall. Beautifully cut, gray, fernlike leaves. Flat-topped clusters of small yellow daisylike flowers in late spring and early summer. Requires well-drained soil, full sun to light shade, little water; cold tender. Propagate from cuttings or seed.

Garden Design Note: Leaves are spectacular. Mass this next to baccharis or coffeeberry for dramatic structure and texture in the garden. Prune to maintain its shape.

Fragrant pitcher sage (*Lepechinia fragrans*).

Shrubs to three feet tall with many horizontally trending branches covered with pairs of evergreen, highly fragrant, softly felted, narrowly oval leaves. Racemes of lavender two-lipped flowers from late winter into spring, followed by pitcherlike, veined sepals that hold four seeds each. Propagate from semihardwood cuttings or seed. Pitcher sage may be cut back by half to keep it more compact.

Garden Design Note: Exquisite sagelike fragrance. Pitcher sage deserves a place beside a frequently traveled path. Mass island yarrow at its base.

TOP: Flowering branches of island bush poppy (*Dendromecon harfordii*). • MIDDLE: Catalina lace (*Eriophyllum nevinii*) foliage and flowers. • BOTTOM: Flowers of island pitcher sage (*Lepechinia fragrans*).

Plants for Rock Outcrops

Santa Cruz Island buckwheat (*Eriogonum arborescens*), ENDEMIC.

Low, broadly mounded shrublet to three feet high with narrow, curled, evergreen leaves and flat-topped clusters of pink to white flowers in summer and fall, fading to rust. Somewhat winter tender. Propagate from seed. Garden Design Note: A very versatile garden plant. Flower display goes on for many months, with many stages of bloom on the plants at the same time, and has great structure when not in bloom. Can substitute for rosemary or lavender. Deer will browse it.

TOP: Santa Cruz Island buckwheat (*Eriogonum arborescens*), lilac verbena (*Verbena lilacina*), and wetern fescue (*Festuca occidentalis*) in a rocky meadow garden. • BOTTOM: Santa Cruz Island buckwheat (*Eriogonum arborescens*) with flowers in several stages.

Giant coreopsis (*Coreopsis gigantea*).
Curious woody perennial with fat trunks to six feet tall, sometimes branched near the top. Large mops of ferny leaves and circles of large, bright yellow daisies in spring. Leaves fall in summer, and trunks remain bare until winter rains return. Propagate from seed.
Garden Design Note: Excellent in a children's garden. This plant cycles through lushness to dormant brown stumplike trunks that, when massed, make you feel like you are walking in a miniature, primordial forest. You may, however, wish to plant fast-growing annuals or ferns to hide the dead-looking stalks.

Prostrate chamise (*Adenostoma fasciculatum* var. *prostratum*), ENDEMIC.
Low, mounded, woody ground cover to three feet tall covered with bright green, evergreen, needlelike leaves. Narrow panicles of tiny white flowers in late spring and summer are followed by rust-colored seed pods. Superior shape compared to the ordinary shrub form of chamise. Propagate from layered branches.
Garden Design Note: Very useful as a rich green textured ground cover in hot climates. Looks great on slopes. Mass with groups of dwarf baccharis.

Island goldenbush (*Hazardia detonsa*); ENDEMIC.
Shrubs to three feet tall with broadly spoon-shaped, silvery gray leaves and narrow spikes of small yellow flower heads of disk flowers only. Outstanding foliage. Flowers in late summer and fall. Propagate from seed.
Garden Design Note: Difficult to find. Has great potential as a late-blooming large perennial.

Dudleya (*Dudleya candelabrum*, *D. greenei*, and others), ENDEMIC.
Bold rosettes of fleshy leaves—*D. candelabrum* with green leaves, *D. greenei* with gray leaves. Candelabralike flowering stalks to a foot high carry many pale yellow flowers in early summer. Propagate from offsets or seed. Old clumps of dudleyas sometimes develop unsightly woody stems underneath the leaf rosettes. When this occurs, cut off the stems and start new plants from these cuttings.

LEFT: Mixed dudleyas. • RIGHT: *Dudleya hassei* (endemic to Santa Catalina Island) as a ground cover. Photograph by S. Ingram.

Garden Design Note: Dudleyas are effective massed in large container gardens with rocks and pebbles of interesting colors and textures.

Sand Dune Plants

Beach goldenbush (*Isocoma menziesii* var. *sedoides*), ENDEMIC.
Prostrate, circular mats up to two feet across. Sticky, bright green, elliptical leaves and a plethora of densely clustered, bright yellow flower heads in summer. Propagate from rooted stems or seed.
Garden Design Note: Can be found in specialty nurseries in southern California.

Rose buckwheat (*Eriogonum grande* var. *rubescens*), ENDEMIC.
Basal clusters of grayish, spoon-shaped leaves form clumps to two feet across and 12 inches high. Long succession of pale pink to ruby red flowers on naked stalks to two feet high in summer and early fall. Propagate from semihardwood cuttings or seed.
Garden Design Note: This is a very showy species when massed on a slope or on mounds. Needs pruning to maintain a tidy appearance. Mix with other buckwheats and accent grasses.

TOP: Rose buckwheat (*Eriogonum grande* var. *rubescens*). Photograph by S. Ingram. • BOTTOM: Rose buckwheat.

Beach suncups (*Camissonia cheiranthifolia*).

Sprawling, short-lived perennial creating mats to two feet across with elliptical, grayish leaves and a long succession of bright yellow, saucer-shaped flowers in late spring and summer. Adaptable to a variety of soils. Propagate from seed. Needs to be restarted frequently.

Garden Design Note: Plant as an annual at the edge of a sunny border. Available in specialty nurseries in southern California.

Additional shade plants to try:

Southern hawkweed (*Hieracium argutum*); white globe-tulip (*Calochortus albus*); common wood fern (*Dryopteris arguta*); creeping snowberry (*Symphoricarpos mollis*); wild sweet pea (*Lathyrus vestitus*); island gooseberry (*Ribes thacherianum*), ENDEMIC; Catalina perfume (*R. viburnifolium*), ENDEMIC; California maidenhair fern (*Adiantum jordanii*); Plummer's baccharis (*Baccharis plummerae*).

Additional chaparral plants to try:

Large buckwheat (*Eriogonum grande*), ENDEMIC; shrub chamise (*Adenostoma fasciculatum*); island redberry buckthorn (*Rhamnus pirifolia*), ENDEMIC; sugar bush (*Rhus ovata*); lemonade berry (*R. integrifolia*); toyon (*Heteromeles arbutifolia*); island scrub oak (*Quercus pacifica*), ENDEMIC.

Additional rock outcrop plants to try:

California fuchsia (*Epilobium canum*), Cleveland lip fern (*Cheilanthes clevelandii*), coffee fern (*Pellaea andromedifolia*), telegraph plant (*Heterotheca grandiflora*), bush senecio (*Senecio flaccidus*), chicory-leaved wandflower (*Stephanomeria cichoriacea*).

Additional sand dune plants to try:

Pink sand verbena (*Abronia umbellata*), southern sand verbena (*A. maritima*), dune bursage (*Ambrosia chamissonis*).

Places to Visit

A visit to one of the major Channel Islands will go far to inspire informed garden design. Regular trips to Santa Catalina Island are available from the Los Angeles area, but our personal favorite is Santa Cruz Island. Concessionaires for boat travel offer access from Santa Barbara and Ventura harbors to Prisoners Harbor and Scorpion Cove, but there are also special tours to the research station run by the Marine Sciences Department at UC Santa Barbara. The rugged topography provides many habitats, and plant communities include chaparral, ironwood groves, closed-cone pine forest, oak woodland, grassland, riparian woodland, coastal sage scrub, and dry arroyos.

nothus (*Ceanothus arboreus*), island paintbrush (*Castilleja lanata* var. *hololeuca*), Santa Cruz Island buck-wheat (*Eriogonum arborescens*), rose buckwheat (*Eriogonum grande* var. *rubescens*), island bush poppy (*Den-dromecon harfordii*), island redberry (*Rhamnus pirifolia*), island mountain mahogany (*Cerco-carpus betuloides* var. *blanchae*), dudleyas (*Dud-leya candelabrum* and *D. greenei*), and island blue-witch (*Solanum clokeyi*).

Special plants endemic to the islands occur in all habitats. Among them are island alumroot (*Heuchera maxima*), fern-leaf ironwood (*Lyonothamnus floribundus* var. *aspleniifolius*), island manzanita (*Arctostaphylos insularis*), tree cea-

Coastal bluffs are studded with all manner of succulents, Dr. Seuss-like tree coreopsis (*Coreopsis gigantea*), and gray-leaved island golden-bush (*Hazardia detonsa*), while protected canyons harbor fern grottoes decorated with monkeyflowers and shaded by majestic island oaks (*Quercus tomen-tella*) and fern-leaf ironwoods. Windswept promontories are adorned with mats of island scrub oak (*Q. pacifica*) and prostrate chamise (*Adenostoma fasciculatum* var. *prostratum*); dunes are illuminated by masses of beach goldenbush (*Isocoma menziesii* var. *sedoides*) and the pinks and reds of rose buckwheat. Chaparral shrubs glow with fruits of many descriptions or with colorful spring blossoms.

BOTTOM: Fraser Point, Santa Cruz Island. • TOP: Flowering plant of *Dudleya candelabrum* on rock face.

CHAPTER 5

DESERT GARDENS: JUXTAPOSING PLANTS FROM AN EXTREME HABITAT

To many, deserts are vast wastelands of barren habitat with no redeeming value. But deserts are deceptive. Look closely, and you will see an amazing array of shrubs, cacti, and small trees, which erupt into riotous color in spring, while in years of generous rains magic carpets of annual wildflowers delight the eye with all the hues of the rainbow.

In these summer-hot, arid habitats, where rains seldom add more than a few inches a year, nothing is more striking than a desert oasis. Desert oases provide hospitable islands of green in an otherwise scorched landscape. They also lend a tropical feel, for the naturally occurring oases that mark the permanent seeps and fault fractures in Joshua Tree National Park, the area around Palm Springs, and Anza-Borrego State Park are immediately recognized by their soaring fan palms (*Washingtonia filifera*). Add a few large, deciduous, broadleaf trees, some smaller, desert-wash trees and shrubs, and an understory of ferns, orchids, and water-loving perennials, and the scene is complete.

LEFT: Borrego Palm Canyon oasis, Anza Borrego State Park • ABOVE: Backlit teddy bear chollas (*Opuntia bigelovii*) and ocotillos (*Fouquieria splendens*) in the early morning light at Anza Borrego State Park, San Diego County.

Intriguing juxtapositions include spiny acacias and mesquites (*Prosopis* spp.), green-barked palo verde (*Cercidium floridum*), muscular Fremont cottonwood (*Populus fremontii*) and western sycamore (*Platanus racemosa*), straight-trunked fan palms, and the exotic blooms of desert-willow (*Chilopsis linearis*) and chuparosa (*Justicia californica*).

Oases are also a strong draw to desert wildlife for water and shade. Hummingbirds flit in and out of shadows, while bighorn sheep, coyotes, and other shy creatures vie for a turn to drink.

Because desert oases are at low elevations—between sea level and 1,500 feet—they are associated with warm deserts that seldom freeze in winter. As in other desert environments, rainfall is less than ten inches annually, often much less. Because of the permanent high water table, which allows water to seep from the ground or run just beneath the soil's surface, life is well supported.

These watering holes are bounded by steep, rocky canyon walls and aprons of rock scree, sand, and silt. Here the drama of the desert is abrupt, with xeric shrubs, cacti, and annual wildflowers replacing the oasis trees and shrubs. A kaleidoscope of color is unleashed from late February to April when rains have been generous: shrubs and cacti vie with the wildflowers for the most vivid hues. A great diversity of flower design reveals the diversity of pollinators that appear in hordes in these years of abundance.

Intermediate deserts—those lying between 2,500 and 4,000 feet in elevation—have yet different mixes of fascinating plants. Here, spiky Joshua trees (*Yucca brevifolia*) and Spanish daggers (*Y. schidigera*) grow with other xeric shrubs and wildflowers.

Where deserts are high and cold with a snow cover in winter, pinyon pines (*Pinus edulis, P. quadrifolia,* and *P. monophylla*) and junipers (*Juniperus* spp.) are interwoven with vast swaths of fragrant gray sagebrush (*Artemisia tridentata*), green rabbitbrush (*Chrysothamnus nauseosus*), dark green antelope brush (*Purshia tridentata*), and others. The growing season of these cold deserts is attenuated, occurring mainly in late spring

Masses of golden monkeyflower (*Mimulus guttatus*) in Redrock Canyon.

and early summer. The highest of our desert mountains support a strange open forest of contorted, long-lived bristlecone pines (*Pinus longaeva*). Although annual wildflowers play a scant role here, there is a plethora of cold-adapted cacti, small flowering shrubs, and showy perennials, which lend themselves to beautiful garden designs.

Creating a Garden

Even a small garden can achieve the welcoming ambiance of an oasis by featuring a single tree near a small water feature—fountain, pond, or basin—with plantings restricted to the tree's shade and the water's edge. As with other wooded places, plant material can be layered, with an overstory tree, a short understory tree or large shrub,

Anza-Borrego Visitors Center stairway to living roof garden.

and herbaceous plants close to the ground. Seasonality is strong here, and many plants take a winter rest.

In a larger desert garden the elements of the oasis—shade and water—become the heart of a rocky, sere landscape: a place of contrasts. Rock features combine beautifully with a wide variety of succulents, desert shrubs, and annuals. Here, the succulents create the framework because they're permanent; the shrubs fill in with colorful spring and early-summer blooms, then go dormant; while the annuals, planted in broad drifts, provide a visual feast in spring but quickly seed and die when conditions turn dry and hot.

Few gardeners will opt for a desert garden featuring shrubs and perennials from the high, cold deserts, although some of these adapt surprisingly well to desert gardens at low elevations.

Gardeners in the inner foothills can achieve a measure of success with a desert garden design if they live in the hot-summer zone beyond the influence of coastal fog. Beware of overly cold, wet winters, though: desert plants will freeze and rot, even if they're happy the rest of the year. Particularly susceptible shrubs and cacti should be grown together and, in winter, covered with polyethylene tents to hold in warmth and keep out excess rain.

Design Notes

This one-acre garden in a gated community in the Fremont hills comprises a series of themed gardens. To develop the gardens, I considered various factors: the site, including directional exposure, slope, soils, and relationship to the house; the owner's tastes and interests; long-term use of the proposed gardens; and budget. All of the areas to be landscaped were disturbed. When the house was under construction, there was extensive grading, compaction, and fill. When I first visited the site, the hardscape and pool were already in place, and invasive weeds were growing out of control.

The southwest-oriented hillside, we determined, was the ideal location for a home vineyard, which we surrounded with a flagstone terrace offering superb views. The vineyard is underplanted with a wildflower meadow and native bunchgrasses.

The lands around the vineyard are a mosaic of chaparral plantings and bunchgrass meadows that bloom with annuals and perennials eight months of the year.

On the northeast-facing slope one sees when approaching the house, we planted a

Desert earth shelter with brittlebush (*Encelia farinosa*) and ocotillo (*Fouquieria splendens*).

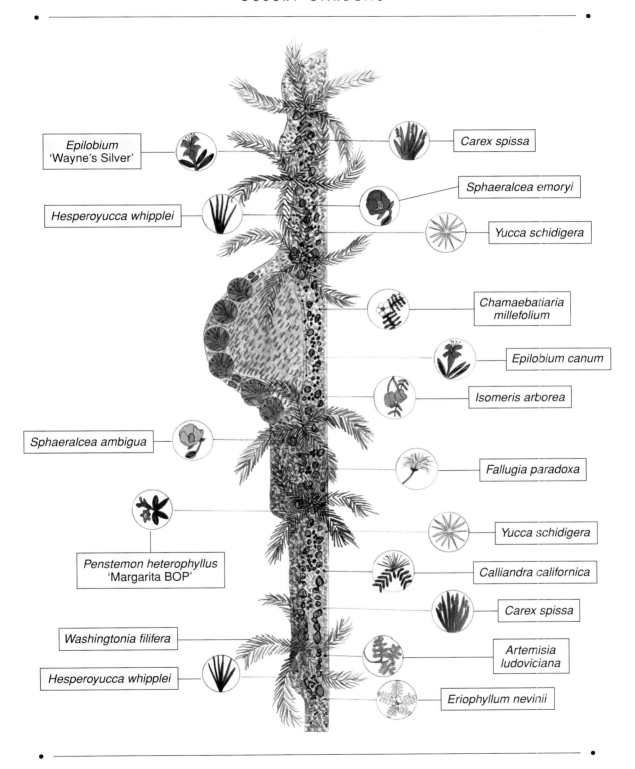

Epilobium 'Wayne's Silver'

Hesperoyucca whipplei

Carex spissa

Sphaeralcea emoryi

Yucca schidigera

Chamaebatiaria millefolium

Epilobium canum

Isomeris arborea

Sphaeralcea ambigua

Fallugia paradoxa

Penstemon heterophyllus 'Margarita BOP'

Yucca schidigera

Calliandra californica

Carex spissa

Washingtonia filifera

Hesperoyucca whipplei

Artemisia ludoviciana

Eriophyllum nevinii

Rendering of featured garden.

mixed-evergreen woodland garden. The owner's desire for privacy influenced a decision to utilize coast live oak, black oak, bay, and madrone. The area beneath these trees is planted with snowberry, creeping barberry, coffeeberry, and toyon.

The formal entry features a courtyard with built-in stucco and brick planters and a rectangular cut bluestone floor. The owner grows tropical plants, which were recycled into a tropical garden. We added bamboo, sago palms, giant bird of paradise, pygmy date palms, and lady palms for size and structure. Because the house provides some protection from winter cold, the plants are thriving.

For the pool area, the owner preferred palms but also desired a California native garden. We created a desert oasis complete with a rocky arroyo and the three palm

TOP: San Diego sedge (*Carex spissa*) and sand verbena (*Abronia villosa*) in the featured garden. • BOTTOM: Hillside vineyard overlooking desert oasis garden.

species native to California and northern Baja. The budget allowed us to crane-in-place six mature palms, each weighing several thousand pounds. We also used the crane to install large boulders. Adjacent to the arroyo is a small wildflower meadow filled with red fescue grass, annuals, and perennials. Plant selections came from among plants commonly found in oases, where the water table is high, as well as desert scrub areas, which receive less than ten inches of rainfall each year. The irrigation throughout is drip. The small meadow is watered with a combination of micromisters and drip lines. Water delivery to each plant can be adjusted to individual moisture requirements.

We also built a small turf area for sunbathing. (The owner is considering replacing this turf area with replicated grass.) It is bordered on one edge by a showy desert oasis grass, alkali sacaton, and on the opposite edge by a series of selections of California

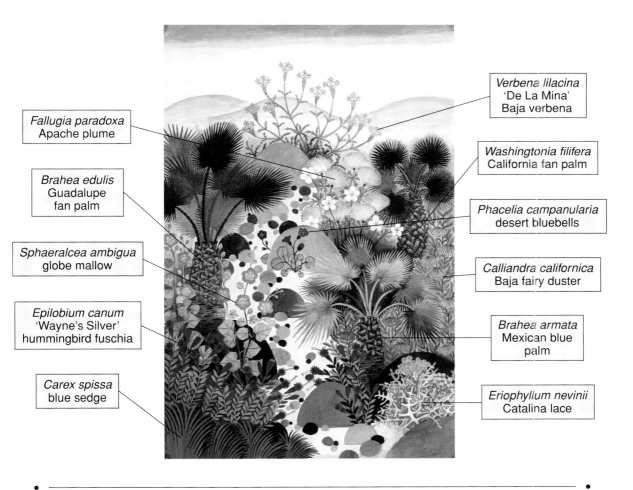

Verbena lilacina 'De La Mina' Baja verbena

Fallugia paradoxa Apache plume

Washingtonia filifera California fan palm

Brahea edulis Guadalupe fan palm

Phacelia campanularia desert bluebells

Sphaeralcea ambigua globe mallow

Calliandra californica Baja fairy duster

Epilobium canum 'Wayne's Silver' hummingbird fuschia

Brahea armata Mexican blue palm

Carex spissa blue sedge

Eriophyllum nevinii Catalina lace

Drawing of featured garden includes hummingbird fuchsia (*Epilobium canum*), globe mallow (*Sphaeralcea ambigua*), Apache plume (*Fallugia paradoxa*), three California palms, and fairy duster (*Calliandra californica*). Illustration by A. Yankellow.

fuchsia. The wall planters behind the pool are filled with colorful desert scrub plantings, the colors and textures of which create interest all year.

The plants selected for this garden, many of which are from northern Baja California and the Channel Islands, are not readily available in northern California. Most were purchased from southern California nurseries and botanic gardens, including Santa Barbara and Rancho Santa Ana botanic gardens, Theodore Payne Foundation, Tree of Life Nursery, Matilija Nursery, and Las Pilitas Nursery, and at California Native Plant Society plant sales.

One of my favorite plants for the desert garden is Catalina lace. Its leaves, with their intricate pattern and texture, formed the inspiration for a mosaic basin that Ananda Yankellow and Christina Yaconelli created for the upper meadow patio.

Every time I visit this desert garden, something is in bloom and I'm eternally surprised by the subtle relationships that occur in this ecosystem.

Scope of Work

Large Palms

- ◆ Build drainage system and create sandy soil beds.
- ◆ Install trees and boulders from overhead with a large truck-mounted crane. (It was parked on the street, since this was our only access to the site.)

Arroyo

- ◆ Build arroyo two to three feet deep by three to four feet wide.
- ◆ Accent with a combination of river-washed black and gray cobbles and Baja pebbles of varying sizes. Add an assortment of other boulders, including some that are blunt and jagged and some larger water-washed rocks.

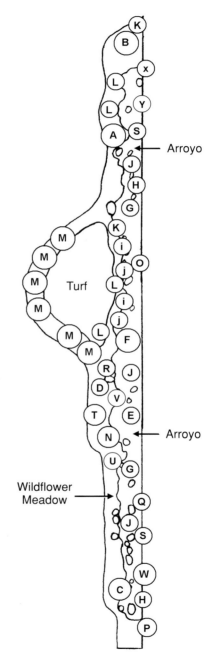

Featured garden key.

Upper Meadow

- Install irrigation.
- Install plants; seed meadow with grass seed and plugs and with wildflower seed.
- Install small turf area.
- Mulch planted areas with gravel.
- Mulch in wall planters with a combination of rocky gravel, sand, and mixed cobbles, as well as small rocks and remnants from the bluestone courtyard construction.

Plant List

Symbol	Botanical Name	Common Name
A	*Washingtonia filifera*	California fan palm
B	*Brahea edulis*	Guadalupe fan palm
C	*B. armata*	Mexican blue palm
D	*Sphaeralcea ambigua*	globe mallow
E	*Fallugia paradoxa*	Apache plume
F	*Isomeris arborea*	bladderpod
G	*Yucca schidigera*	Spanish dagger
H	*Hesperoyucca whipplei*	chaparral yucca
i	*Epilobium canum*	California fuchsia
j	*E. canum*	hoary form of California fuchsia
K	*E. 'Catalina'*	Catalina fuchsia
L	*E. 'Wayne's Silver'*	silver Californica fuchsia
M	*Sporobolus airoides*	alkali sacaton
N	*Adiantum capillus-veneris*	southern maidenhair
O	*Chamaebatiaria millefolium*	fern bush
P	*Eriophyllum nevinii*	Catalina lace
Q	*Calliandra californica*	Baja fairy duster
R	*Galvezia juncea*	Baja bush snapdragon
S	*Verbena lilacina*	lilac verbena
T	*Penstemon palmeri*	Palmer's penstemon
U	*Eschscholzia glyptosperma*	desert California poppy
V	*Juncus patens*	California rush
W	*Artemisia ludoviciana*	silver wormwood
x	*Tecoma stans*	yellow bells
Y	*Sphaeralcea emoryi*	globe mallow

Plants to Use

Large Trees

California fan palm
(*Washingtonia filifera*),
OASIS.

Closely related to the widely planted Mexican fan palm (*W. robusta*), California fan palm is stouter and shorter, reaching a height of 50 feet in age. Large mops of broadly fan-shaped leaves top each tree, but the old fronds are retained as a skirt, lending these trees a special character. Hundreds of small white flowers in hanging panicles in late spring are followed by small, sweet, datelike fruits in fall. Propagate from seed. Little maintenance needed; retaining the old thatch of leaves adds to the drama of this plant.

Garden Design Note: Young plants work well in containers for tropical effect. Elevate them, to create a canopy around a patio.

Mexican blue palm
(*Brahea armata*).
Single-trunked evergreen palm tree to 35 feet tall. Huge fan-shaped blue, palmately lobed leaves. Long, drooping panicles of small white flowers appear in late spring to early summer; small datelike fruits follow. Requires well-drained soil, full sun, some water. Propagate from seed. Little maintenance needed. Native to Baja California.

Garden Design Note: Beautiful blue palms can be featured in a blue desert oasis garden together with lupines, desert bluebells and 'Canyon Prince' grass.

TOP: California fan palm (*Washingtonia filifera*) and *Leymus condensatus* 'Canyon Prince'. • BOTTOM: Mexican blue palm (*Brahea armata*) with pink alum root (*Heuchera* 'Wendy').

Fremont cottonwood (*Populus fremontii* ssp. *fremontii*), OASIS.
Deciduous, fast-growing tree to 50 feet high with spreading crown. Fissured bark and broadly triangular, fresh green, toothed leaves that turn yellow in fall. Trees are male or female: males have drooping catkins of wind-pollinated flowers; females have cat-kins of pistillate flowers that become capsules with hundreds of cottony, white, hair-covered seeds. Avoid female trees because of the mess created by this cotton. Propa-gate from cuttings or suckers to ensure the sex of the tree. Extra suckers should also be removed as they appear, unless a grove of cottonwoods is wanted.
Garden Design Note: Requires lots of space. Can be shaped artistically when young to provide structure for a shady enclosure.

Short Trees

Palo verde (*Cercidium floridum*), OASIS.
Small, often multitrunked tree to 20 feet high with distinctive green bark. Spiny branches carry small, pealike leaves after rains. Tree crowns are transformed in midspring by masses of bright yellow, butterflylike flowers followed by pealike seed pods. Propagate from scarified seed. The old seed pods and flowers can be al-lowed to form a natural mulch.
Garden Design Note: Very showy small

flowering tree for the desert landscape. Several planted in an open garden setting make an effective display.

Smoke tree (*Psorothamnus spinosus*), OASIS.
Distinctive, leafless, rounded tree to 15 feet high from washes. Intricate, curved, gray-ish green twigs carry on photosynthesis year round. Trees look like large puffs of gray smoke. Hundreds of deep purple blossoms appear in late spring, followed by short, pealike seed pods. Propagate from scarified seed. Little maintenance needed.
Garden Design Note: Although this tree has great potential for the garden, it is diffi-cult to locate in the trade.

Desert-willow (*Chilopsis linearis*), OASIS.
Fast-growing, deciduous, multitrunked tree to 20 feet high with purplish new bark and gray older bark. Narrow, lance-shaped, willowlike leaves sprout in midspring, followed by racemes of gorgeous pale to deep pink, tubular, orchidlike flowers in summer. Select plants in flower for best color forms. Propagate from layered branches or seed. Culti-

Crown of palo verde (*Cercidium floridum*) in peak bloom.

vars available. Thinning branches from time to time can improve the form of the tree. Garden Design Note: This flowering tree can be trained to form a ramada-like structure in the desert garden, or prune it into a multibranched small tree.

Shrubs

Desert-lavender (*Hyptis emoryi*), OASIS.
Fast-growing, tall, slender shrubs to 10 feet high with pairs of pale green, fragrant, ovate leaves and slender spikes of pale, lavender-scented, purple flowers. Each small flower is surrounded by woolly sepals. Long bloom time from winter to spring. Propagate from suckers, cuttings, or seed.

Chuparosa (*Justicia californica*), OASIS.
Multibranched shrubs to three feet tall with many horizontally arching, greenish stems. Ephemeral leaves appear after rains, with racemes of vivid scarlet, wide two-lipped flowers emerging from winter to spring. Attractive to hummingbirds. Propagate from cuttings.

Desert-olive (*Forestiera pubescens*), OASIS.
Densely branched, deciduous shrub to 10 feet high with thorny side branches and bright green, narrowly elliptical leaves. Tiny clusters of fragrant, greenish male or female flowers and small, purple, olivelike fruits. Propagate from scarified seed or suckers. Thin dense twigs from time to time to improve air flow to the crown.
Garden Design Note: Excellent as a hedge or barrier.

Globe mallow (*Sphaeralcea ambigua*).
Shrublets with many upright branches to

TOP: Chuparosa (*Justicia californica*) in flower. • BOTTOM: Large flowering plant of globe mallow (*Sphaeralcea ambigua*).

three feet tall carrying wrinkled, round, gray-green, lobed leaves and slender wands of pale orange to red-orange to wine red, hollyhocklike flowers of great beauty. Long blooming with some water; may rebloom in fall. Propagate from suckers, semihardwood cuttings, or seed. Every year or so, cut out old stems that no longer produce viable growth.

Garden Design Note: Because the flowers of this shrub have such a wide color range, when planted together they create dazzling displays.

Big sagebrush (*Artemisia tridentata*).

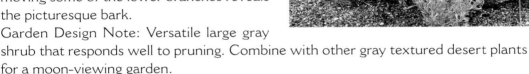

Rounded shrub to eight feet high, relatively fast growing. Fibrous, decorative bark on old trunks; evergreen, vertically oriented, triangular, sage-scented, gray leaves; narrow spikes of greenish yellow, wind-pollinated flowers in fall. Propagate from stratified seed or cuttings. Sagebrush can be severely sheared back to resprout, or left intact. Removing some of the lower branches reveals the picturesque bark.

Garden Design Note: Versatile large gray shrub that responds well to pruning. Combine with other gray textured desert plants for a moon-viewing garden.

Blue desert sage (*Salvia dorrii*).

Low, mounded shrub to two feet high with broad, silvery gray, strongly scented leaves and dense whorls of vivid blue, two-lipped flowers. Flowers are often complemented by pinkish or rose-colored bracts. Propagate from seed or semihardwood cuttings. Needs perfect drainage and resents excessive winter rain.

Garden Design Note: Like many desert species, the strong fragrance of this plant commands a place in a scented garden. Use with Palmer's penstemon.

Apache plume (*Fallugia paradoxa*).

Small, densely branched shrubs to five feet high with semievergreen, pinnately lobed leaves, flat-topped clusters of white, single roselike flowers in late spring, and puffs of feathery, pink-plumed fruits. Propagate from hardwood cuttings or stratified seed. Periodically prune out old branches to open up the structure of the shrubs.

TOP: Big sagebrush (*Artemisia tridentata*) with budded flower spikes. • BOTTOM: Pink plumed fruits of Apache plume (*Fallugia paradoxa*).

Garden Design Note: Fruiting stage is the most attractive. Mass it for full impact in a California gray and white desert garden.

Brittlebush (*Encelia farinosa*).
Low, multibranched shrubs to three feet high with brittle twigs and dense clusters of broad, silvery white leaves. Summer deciduous. Flat-topped clusters of bright yellow daisies appear from early to midspring. Propagate from hardwood cuttings or seed. Garden Design Note: Brittlebush is one of the dominant shrubs in southern California deserts. I have not planted it in northern California.

Succulents

Spanish dagger or Mojave yucca (*Yucca schidigera*).
Mostly single-trunked shrub to 15 feet high, sometimes branched above. Long, sword-

shaped, daggerlike leaves are lined with curled threads, with massive clusters of waxy, white, bell-shaped flowers appearing several feet above the leaves. Propagate from seed. Although the old leaves can be periodically removed, they're difficult to compost because of their dense fibers.
Garden Design Note: Plant away from garden paths and give it lots of room. Leaves stay on the plant to the ground when young. Will bloom after eight years. Loves a slope.

Desert agave (*Agave deserti*).
Broad rosettes of fleshy, ovate, bluish green leaves lined with recurved spines and tipped by a daggerlike needle. Fat, asparaguslike flower buds emerge after 10 to 15 years to produce an enormous panicle up to 12 feet high of tubular, yellow flowers. The parent rosette dies after flowering, but not before producing a circle of "pups." Propagate from pups or seed. Dead leaves and flowering stalks are tough and fibrous and not easily composted.
Garden Design Note: Use with other bluish desert species, like Mexican blue palm and thistle sage. Avoid if children or pets are garden visitors.

Barrel cactus (*Ferocactus cylindraceus*).
Slow-growing, barrel-shaped stem, seldom branched, to six feet high. Nodes are cov-

Flowering plant of Spanish dagger (*Yucca schidigera*).

ered with flat, recurved, pink spines. A whorl of yellow flowers near the top of the barrel appears in April and May. Propagate from offsets or seed.

Garden Design Note: Plant in mass for dramatic effect with contemporary architecture. Feature in raised architectural planters where drainage and gravelly soil can be controlled.

Beavertail cactus (*Opuntia basilaris*).

Creeping stems produce large colonies of flattened, vertically oriented, grayish pads (often turning rose-purple in winter), topped in midspring by large, multipetaled, pink to rose flowers of great beauty. Nodes carry clusters of tiny spines that are not obvious at first glance. Propagate from stem cuttings or seed. Shrunken and dark old pads should be cut off with a sharp knife and discarded during dry weather.

Garden Design Note: This cactus offers great visual interest year round. Effective with prince's plume or desert poppies. Plant high in gravel; top dress with weedcloth, then cover with two inches of chiseled stone and gravel pieces.

TOP: Pads of beavertail cactus (*Opuntia basilaris*) with flowers. • BOTTOM: Prickly pear (*Opuntia* sp.) in fruit in a raised bed.

Mound cactus (*Echinocereus triglochidiatus*).
Creeping stems create dense colonies of nearly globe-shaped green stems with re-curved whitish spines. Colonies grow in rock crevices. Showy, strawberry red, multi-petaled flowers appear mid to late spring. Propagate from stem cuttings or seed.
Garden Design Note: Another fine species for mass planting amid rocky outcroppings.

Perennials

Seep columbine (*Aquilegia shockleyi* [= *A. formosa* in *The Jepson Manual*]), OASIS.
Winter-dormant perennial from sturdy rootstocks. Ternately compound, grayish green leaves; flower stalks to eighteen inches high with bright red and yellow, spurred flow-ers attractive to hummingbirds. Foliage and stems have sticky glands. Propagate from seed. Remove old leaves at the end of the year.
Garden Design Note: Plant several of these plants beside a desert arroyo in the shade of a fan palm. Needs summer water.

Seep goldenrod (*Solidago confinis*), OASIS.
Winter-dormant perennial from short rhizomes. Basal clumps of narrowly spoon-shaped, dark green leaves and leafy stems carry large, dense panicles of miniature, bright yellow daisies in late summer. Propagate from root divisions or seed. Cut back flowering stalks to the ground in late fall or winter.
Garden Design Note: Adds late summer color to a desert meadow planted with reed grass, alkali sacaton, epilobium, and buckwheats. Give some summer water.

Desert-aster (*Xylorhiza tortifolia*).
Woody-based perennial to three feet high with semievergreen, coarsely toothed leaves and large, single, blue-purple and yellow daisies. Flowers ap-pear in winter and spring. Propagate from semihardwood cuttings or seed.

Prince's plume (*Stanleya* spp.).
Dramatic, subwoody perennial with several wandlike flowering stalks to six feet high from basal clump of simple to pinnately divided blue-green leaves. Bright yellow flowers have protruding stamens and are visited by many pollinators. Propagate from semihardwood cuttings or seed. Cut off old flowering stalks at summer's end.
Garden Design Note: One of my favorite desert species. Plant several, along with Palmer's penstemon, in well-drained gravel, then sprinkle the bed with desert bluebell seeds for a brilliant spring display. Needs a couple of years to become established.

Showy flowers of desert-aster (*Xylorhiza tortifolia*).

Palmer's penstemon (*Penstemon palmeri*).
Subwoody perennial forming low mounds to a foot tall with folded, elliptical, coarsely toothed leaves and spectacular racemes of inflated, pale purple flowers on stalks to six feet high. Powerfully fragrant flowers are decorated with purple lines and a yellow-bearded sterile stamen. Propagate from divisions, cuttings, or seed. Palmer's penstemon seldom lives more than a few years; cuttings should be taken in spring to assure a continued supply of healthy plants.
Garden Design Note: Excellent selection for a gray-themed desert garden. Mass this pink penstemon next to a sunny entrance in gravelly soils to get its heady scent.

Desert marigold (*Baileya multiradiata*).
Basal clusters of pinnately lobed leaves atop a taproot. Flowering stalks to two feet high carry flat-topped clusters of golden yellow daisies with turned-down rays. Propagate from seed.
Garden Design Note: Plant as an annual or short-lived perennial. Use with rocks and gravel mulch.

Angel's trumpet or thorn-apple
(*Datura wrightii*).
Winter-dormant, clumping plant to two feet high with coarse, triangular, ovate leaves and very large, single, trumpet-

TOP: Flowering plant of Palmer's penstemon (*Penstemon palmeri*). • BOTTOM LEFT: Flowering plant of angel's trumpet (*Datura wrightii*). • BOTTOM RIGHT: Desert marigold (*Baileya multiradiata*) in a sunny border.

shaped white or purple-flushed fragrant flowers that open in evening. Propagate from seed.

Garden Design Note: Very showy flowers will arch over a raised planter or spill over edges in container gardens. Flowers and plant are poisonous, so don't use in a children's garden. Effective as an annual because it grows fast. Use around a patio where the full moon can be viewed.

Grasses

Common reed (*Phragmites australis*), OASIS.
Giant grass with bamboolike stems to twelve feet tall from stout rhizomes. Dense plumes of spikelets in summer. Cut back frequently to encourage new growth. Propagate from divisions of rhizomes. Thin clumps every few years to maintain best health.
Garden Design Note: Dramatic grass to complement contemporary architecture.

Alkali sacaton (*Sporobolus airoides*), OASIS.
Bold clumps of stiff, curved, dark green leaves to about a foot high. Airy, much-branched panicles to two feet high carry tiny spikelets in late summer. Tolerates alkaline soils. Propagate from divisions or seed. At year's end, remove and recycle old dead leaves.
Garden Design Note: Mass this ethereal grass under a palm grove and watch a summer wind carry the tiny blooms away.

Ferns

Southern maidenhair (*Adiantum capillus-veneris*), OASIS.
Small, winter-dormant fern to six inches high. Wiry, black stalks and delicate, fanlike segments on fronds, similar to florist's maidenhair fern. Beautiful when massed. Propagate from divisions or spores.
Garden Design Note: Tuck this delicate-looking fern into shady rock crevices and allow fresh water to trickle around it.

Southern water fern (*Thelypteris puberula* var. *sonorensis*), OASIS.
Rhizomatous fern with long tapered fronds eighteen inches to two feet long, each frond twice divided. Propagate from divisions of rhizomes or spores. Cut off old, dried fronds at year's end.
Garden Design Note: If you are lucky enough to find this fern, buy it at once and place it prominently near a water feature where all can see it.

Vines

Desert grape (*Vitis girdiana*), OASIS.
Vigorous, winter-deciduous, fast-growing, woody vine with typical grapelike leaves, clusters of yellow-green flowers attractive to bees, and edible, but not choice, purple fruits. Similar to California grape (*V. californica*), except young leaves are downy. Desert grape can be periodically pruned back to its woody canes to keep its growth in bounds. Garden Design Note: Cover a rustic ramada with this vine, or create vine tracery against a painted stucco wall.

Annuals

Propagation Note: All annuals are propagated from seed.

Chia (*Salvia columbariae*).
Textured, irregularly pinnately lobed, fragrant leaves rise from a basal rosette. Several flowering stalks carry whorls of spiny-bracted, clear blue, two-lipped flowers complemented by red-purple sepals. Grows to three feet tall with ample water. Extra water prolongs the bloom period and size of plants.
Garden Design Note: Excellent in a children's garden. Seeds are nutritious. Plant in

Desert grape (*Vitis girdiana*) on rock wall under a manzanita (*Arctostaphylos* sp.).

open gravelly areas with desert bluebells, tansy leaf phacelia, elegant tarweed, and poppies.

Thistle sage (*Salvia carduacea*).
Bold, grayish, cobwebby-haired, thistlelike leaves in a basal rosette. Stout flower stalks carry several whorls of gorgeous, pale lavender-blue, frilled, two-lipped flowers surrounded by spiny bracts. Choice but difficult.
Garden Design Note: I have never found this sage in cultivation, but am spellbound by it in the wild. Lester Rowntree says that it should be grown in loose gravel.

Desert bluebells (*Phacelia campanularia*).
Rounded, scalloped dark green leaves and branched flower stalks to three feet tall. Long succession of true blue, bell-shaped blossoms of great beauty in midspring. Extra water prolongs the bloom period.
Garden Design Note: We've used this successfully in our hydroseeded meadows, but we must continually reseed, as it doesn't seem to naturalize in northern California.

Desert sand-verbena (*Abronia villosa*).
Widely sprawling, sticky stems bear oval, fleshy leaves and dense, headlike clusters of rose- or pink-purple, verbena-like, fragrant flowers. May bloom in winter where temperatures are mild. Stunning when massed. Prefers sandy soil. Clumps can grow to over three feet across with constant moisture.
Garden Design Note: Short-lived; treat as an annual. Makes a big show with monardella, lilac verbena, and poppies.

Satin blazing star (*Mentzelia involucrata*).
Plants to three feet tall with elliptical, rough-textured leaves, silvery with glistening white stems, and large, cup-shaped, creamy flowers with a satiny texture and subtle pink stripes.
Garden Design Note: Difficult both to find and to grow. Plant with excellent drainage and mass.

Birdcage evening-primrose (*Oenothera deltoides*).
Spreading crowns of leafy stems bear clusters of very large, fragrant, white flowers that open in the evening. Extended bloom period from winter to spring with extra

TOP: Desert bluebells (*Phacelia campanularia*). • BOTTOM: Desert sand-verbena (*Abronia villosa*).

water; clumps may reach two feet across. Spent blooms and stems create a contained dry basket that blows across desert dunes.

Desert-sunflower or desert gold (*Geraea canescens*).
Bushy, annual or short-lived perennial to three feet high with ovate, bright green, toothed leaves and clusters of large, golden yellow daisies from winter to spring. Can hybridize with brittlebush. In especially rainy years, the flowers carpet the desert floor for acres.
Garden Design Note: Difficult to find.

Bottle scrubber (*Camissonia boothii*).
Low-growing annual to eight inches high with rosettes of narrow, red-spotted leaves

Desert sunflower (*Geraea canescens*).

and dense spikes of flowers with red buds and pink to white petals in spring. Skeleton looks like a bottle scrubber.
Garden Design Note: Difficult to find.

Desert coreopsis (*Coreopsis bigelovii*).
Basal rosettes of pinnately divided leaves and several single-flowered stalks with large, intensely gold daisies.
Garden Design Note: Difficult to find.

Desert dandelion (*Malacothrix glabrata*).
Basal rosettes of finely pinnately divided leaves and leafless stalks with large pale yellow dandelionlike flowers. Grows to a foot high.
Garden Design Note: Difficult to find.

Additional trees to try:

Narrow-leaf willow (*Salix exigua*), OASIS; white alder (*Alnus rhombifolia*), OASIS; mesquites (*Prosopis glandulosa* and *P. pubescens*), OASIS; desert ironwood (*Olneya tesota*), OASIS; single-needle pinyon (*Pinus monophylla*); Utah juniper (*Juniperus osteosperma*); California juniper (*J. californica*).

Desert garden featuring desert-willow (*Chilopsis linearis*) and boulders.

Additional shrubs to try:

Arrowweed (*Pluchea sericea*), OASIS; catclaw acacia (*Acacia greggii*), OASIS; false indigo (*Amorpha fruticosa*), OASIS; Wood's rose (*Rosa woodsii*), OASIS; broom baccharis (*Baccharis sergiloides*), OASIS; banana yucca (*Yucca baccata*); silver cholla (*Opuntia echinocarpa*); creosote bush (*Larrea tridentata*); hopsage (*Grayia spinosa*); desert-holly (*Atriplex hymeneletra*); rabbitbrush (*Chrysothamnus nauseosus*); indigo bush (*Psorothamnus fremontii* and *P. schottii*); desert senna (*Senna armata*); paperbag bush (*Salazaria mexicana*); cliff-rose (*Purshia mexicana* var. *stansburyana*).

Additional perennials to try:

California loosestrife (*Lythrum californicum*), OASIS; seep aster (*Aster lanceolatus*), OASIS; wild licorice (*Glycyrrhiza lepidota*), OASIS; yellow spiderflower (*Cleome lutea*); desert-nettle (*Eucnide urens*); prickly poppy (*Argemone munita*); Eaton's firecracker (*Penstemon eatonii*); Utah penstemon (*P. utahensis*).

Additional annuals to try:

Desert monkeyflower (*Mimulus bigelovii*), desert and Parish's poppies (*Eschscholzia glyptosperma* and *E. parishii*), fernleaf phacelia (*Phacelia distans*), royal phacelia (*P. crenulata*), yellow-throat (*P. fremontii*), royal lupine (*Lupinus odoratus*), fragrant gilia (*Gilia latiflora*), desert-chicory (*Rafinesquia neomexicana*), desert five-spot (*Eremalche rotundifolia*).

CLOCKWISE FROM TOP LEFT: Narrowleaf milkweed (*Asclepias fascicularis*). • Desert-willow (*Chilopsis linearis*). • Bigpod mariposa-tulip (*Calochortus macrocarpus*).

Places to Visit

Borrego Palm Canyon, San Diego County. (Low desert with palm oasis.) This popular canyon lies northwest of the visitor's center in Anza-Borrego State Park, the largest of California's state parks, located west of the Salton Sea and east of the Laguna and Cuyamaca mountains. The round-trip hike to the palm oasis is around three miles and features cactus and creosote bush scrub with spectacular rocks and boulders. Visit between late February and early April for

TOP: Desert brittlebush (*Encelia farinosa*) in flower. • BOTTOM: Clouds over mountains, mouth of Borrego Palm Canyon.

the best wildflowers. Trees include California fan palm (*Washingtonia filifera*), catclaw acacia (*Acacia greggii*), honey mesquite (*Prosopis glandulosa*), white alder (*Alnus rhombifolia*), western sycamore (*Platanus racemosa*), and Fremont cottonwood (*Populus fremontii*). Among the many shrubs, highlights include rock hibiscus (*Hibiscus denudatus*), brittlebush (*Encelia farinosa*), chuparosa (*Justicia californica*), desert-lavender (*Hyptis emoryi*), cheesebush (*Hymen-oclea salsola*), creosote bush (*Larrea tridentata*), wand sage (*Salvia vaseyi*), and globe mallow (*Sphaeralcea ambigua*). Cacti include buckhorn cholla (*Opuntia acanthocarpa*), beavertail cactus (*O. basilaris*), barrel cactus (*Ferocactus cylindraceus*), hedgehog cactus (*Echinocereus engelmannii*), and fishhook cactus (*Mamillaria dioica*).

Cottonwood Springs, Joshua Tree National Park, Riverside County. (Low Sonoran desert.) Cottonwood Springs is near the south end of Joshua Tree National Park on Pinto Basin Road. There is a small visitor's center, a campground, some hiking trails, and beautifully sculpted granite rocks. Extensive alluvial fans to the south host some of the best displays of annual wildflowers in the area. Vegetation around the springs includes creosote bush scrub and palm oasis. Just to the north are good stands of ocotillo (*Fouquieria splendens*) and teddy bear cholla (*Opuntia bigelovii*). The local flora includes desert-ironwood (*Olneya tesota*), California fan palm (*Washingtonia filifera*), prince's plume (*Stanleya pinnata*), desert bluebells (*Phacelia campanularia*), bladderpod (*Isomeris arborea*), chuparosa, desert monkeyflower (*Mimulus bigelovii*), brittlebush, indigo bush (*Psorothamnus schottii*), creosote bush, chia (*Salvia columbariae*), suncups (*Camissonia* spp.), birdcage evening-primrose (*Oenothera deltoides*), desert star (*Monoptilon belloides*), and satin blazing star (*Mentzelia involucrata*).

Caruthers Canyon, San Bernardino County. (Intermediate Mojave Desert.) Caruthers Canyon is in the New York Mountains in the Mojave National Reserve at the interface between Joshua tree woodland and pinyon pine forest with a strong component of desert chaparral. To find this canyon, consult a local map; high-clearance vehicles are recommended. From road's end, you can hike three miles to an old mine. Plants of merit include Mexican manzanita (*Arctostaphylos pungens*),

Granite rocks with juniper and pinyon pine, Joshua Tree National Park.

desert silk tassel (*Garrya flavescens*), narrowleaf yerba santa (*Eriodictyon angustifolium*), desert scrub oak (*Quercus turbinella*), desert ash (*Fraxinus anomala*), Utah firecracker penstemon (*Penstemon utahensis*), Mojave thistle (*Cirsium mohavense*), bladderpod (*Lepidium kingii*), desert phlox (*Phlox stansburyi*), several cacti, desert verbena (*Verbena goodingii*), and Spanish dagger (*Yucca schidigera*).

Wildrose Canyon, Death Valley National Park, Inyo County. (High desert.) Wildrose Canyon is in the Panamint Mountains on the western side of Death Valley between Panamint Valley and Death Valley proper. An access road from Hwy 178 leads to old charcoal kilns and a campground at Mahogany Flats. The drive from Panamint Valley to Mahogany Flats passes through shadscale scrub, creosote bush scrub, desert riparian oasis, mixed cactus scrub, sagebrush scrub, blackbrush scrub, and pinyon juniper woodland, as well as spectacular limestone cliffs. Plant highlights include Panamint daisy (*Enceliopsis covillei*), Fremont cottonwood, Wood's rose (*Rosa woodsii* var. *ultramontana*), prince's plume, ghost flower (*Mohavea breviflora*), desert monkeyflower, hopsage (*Grayia spinosa*), blue desert sage (*Salvia dorrii*), goldenhead (*Acamtopappus shockleyi*), desert nettle (*Eucnide urens*), cottontop cactus (*Echinocactus polycephalus*), old man's prickly pear (*Opuntia erinacea*), mound cactus (*Echinocereus triglochidiatus*), desert mariposa

(*Calochortus kennedyi*), Death Valley penstemon (*Penstemon fruticiformis*), magnificent lupine (*Lupinus magnificus*), cliff-rose (*Purshia mexicana* var. *stansburyana*), and curl-leaf mountain mahogany (*Cercocarpus ledifolius*).

Grandview Camp, White Mountains, Inyo County. (High desert mountains in the Great Basin.) Take Hwy 168 east from Hwy 395 near Big Pine (south of Bishop in the Owen's Valley). Just before Westgard Pass, turn north onto White Mountain Road, go eight miles, and turn left into Grandview campground. This area is just a couple of miles below the famous bristlecone pines, oldest living trees, and is a fine example of high desert vegetation. The two major plant communities are pinyon-juniper woodland and sagebrush scrub. Plants include blue desert sage, big sagebrush (*Artemisia tridentata*), cushion buckwheat (*Eriogonum caespitosum*), White Mountains penstemon (*Penstemon scapoides*), showy penstemon (*P. speciosus*), scarlet mountain penstemon (*P. rostriflorus*), single-needle pinyon (*Pinus monophylla*), Utah juniper (*Juniperus osteosperma*), cliff-rose, old man prickly pear, Mormon tea (*Ephedra* spp.), desert phlox, Coville's phlox (*Phlox condensatus*), soda straw angelica (*Angelica lineariloba*), and green gentian (*Swertia puberulenta*). Beautiful desert plants along the road up include scarlet locoweed (*Astragalus coccineus*), prickly poppy (*Argemone munita*), angel's trumpet (*Datura wrightii*), common blazing star (*Mentzelia laevicaulis*), rose penstemon (*Penstemon floridus*), wedgeleaf goldenbush (*Ericameria cuneata*), Heerman's buckwheat (*Eriogonum heermanii*), and flattop buckwheat (*E. fasciculatum*).

Barrel cactus (*Ferocactus cylindraceus*).

Palm oasis in Hellhole Canyon, Anza Borrego State Park, San Diego County.

CHAPTER 6

MONTANE MEADOWS: GARDENING WITH MOUNTAIN WILDFLOWERS

You know you're in the heart of the mountains when you see grassy meadows ablaze with perennial wildflowers along dashing streams or encircling quiet lakes. Montane meadows provide visual relief from the great bands of mixed conifer forest at elevations of 4,000 feet to above timberline. Magic tapestries of summer wildflow-

ers bloom among a permanent framework of sedges and perennial bunchgrasses. Starting sometime in June, just after snowmelt, the floral pageant continues through the summer, ending only when meadows turn dry. The progression of bloom draws a wide variety of pollinators, ranging from tiny native bees and bee-mimicking hoverflies to colorfully winged butterflies and darting hummingbirds.

Meadows are temporary landscapes in the overall scheme of nature's habitats. They occupy receding ponds and lakes as the water bodies gradually silt up; with the accumulation of more soil, willows and alders move in, followed by the encroaching conifer forest. Many montane meadows were maintained in their grassy state by Native Americans through the use of fire; without burning, the meadows gradually turn into forests. Yosemite Valley was once a parkland of stately conifers fringing open meadows, but today a different scenario is evolving as brushy, overgrown forests take over shrinking meadows. Difficult management decisions await park personnel.

Meadow species vary according to elevation and moisture regime, which ranges from soggy to dry. In the garden, you can take advantage of this diversity by creating a multilevel meadow with varying water requirements. Within the framework of a montane meadow, the wet parts feature sedges (*Carex* spp.), with drier areas given over to a mixture of perennial bunchgrasses. Perennial wildflowers and bulbs will fill in the gaps, so be sure to allow ample room for them in your garden design.

LEFT: Blue elderberry (*Sambucus mexicana*) and sagebrush (*Artemisia tridentata*) in the high eastern Sierra Nevada. •
ABOVE: Lupines (*Lupinus albicaulis*) and meadow paintbrush (*Castilleja miniata*), Lake Winnemucca.

The climate of a montane meadow matches that of the surrounding forest: snowy winters and warm summers with occasional rain. Subalpine meadows experience longer, colder winters with extended dormancies and a shorter growing season than meadows at middle elevations. At the highest elevations, summers are cool with frequent thunderstorms. Choose plant material based on these patterns. The soils of most meadows are deep and rich with organic material in the form of peat.

TOP: Red-heather (*Phyllodoce breweri*) at Lake Winnemucca. • BOTTOM LEFT: Meadow of mixed wildflowers near Lake Winnemucca. • BOTTOM RIGHT: Common corn-lily (*Veratrum californicum*) in a montane meadow.

Creating a Garden

Although lower-elevation montane meadows can be recreated with modest success in foothill gardens, they are most likely to succeed around mountain cabins or in mountain towns. Unlike their pine-belt counterpart, meadows need full sun and appreciate ample moisture. Soils should be slightly acid.

Be sure to use a combination of short and tall perennials, and choose species that span the bloom period from snowmelt to summer's end. Many meadow plants also provide drama in the garden in their fruiting state or when their leaves turn color in fall, and grasses often later dry to beautiful forms.

TOP TO BOTTOM: Fawn lily (*Erythronium klamathense*). • Alpine lily (*Lilium parvum*). • Villa Gigli, Markleeville, California.

Dry, scree area

Jeffrey pine, white fir and incense cedar forest

Villa Gigli

Mountain stream

Mountain meadow

Wooden plank bridge

Gigli ceramic sculpture collection

Outdoor dining area

Split rail cedar fence

Gravel drive and parking lot

Design Notes

The site for this mountain meadow garden is near the town of Markleeville on the Grover Hot Springs Road at an elevation of approximately 5,500 feet. Its design was inspired by a wildflower hike in the Sierra near Winnemucca Lake during mid-July—typically the best time of year to see mountain meadows at their peak. The building in the design, Villa Gigli, is the home, gallery, and mountain trattoria of Gina and Ruggero Gigli. The design, however, exists only as an artistic vision.

To create a mountain meadow of this diversity, the following environmental conditions are necessary: a high elevation—3,500 to 8,000 feet; abundant water, preferably near a mountain stream; loose, crumbly soils; good drainage; and full sun for eight to 10 hours. Plants are grouped according to their water needs. Many species, such as mountain iris and cotton grass, will colonize, so adequate space must be allowed for expansive monocultures. Tall species like lupine, delphinium, and monkshood, planted near the creek, define the upward reaches of the garden. In rocky areas where it is drier, species like mule's ear, sulfur buckwheat, and scarlet gilia do well.

Important flowering shrubs that give structure and bold color to this garden are the mountain elderberry, planted as accents along the fence. Deerbrush and greenleaf

Mountain meadow concept garden.

manzanita are effective as background species, or massed as understory plantings at the edge of the red and white fir forest.

The more open meadow areas feature mountain grasses and sedges such as Berkeley sedge, bent grass, mountain muhly, and mountain oat grass. These grassy areas provide a quiet contrast to the riotous colors of the meadow's peak weeks, which extend from June to August.

Several species in the design are not included in the plant list, generally because they are not commonly available. I expanded the list, however, so if some plants can't be found, you will have an idea of others that will work as substitutes.

Scope of Work

- ◆ Site garden: it should be on a slope.
- ◆ Assess water availability. Is there an existing mountain stream or good seasonal runoff? Will a water feature be needed with a recycling pump to provide additional irrigation to the meadow area?
- ◆ Plant and group species according to water requirements.
- ◆ Install irrigation according to watering zones.
- ◆ Add boulders.
- ◆ Plant and seed meadows.
- ◆ Mulch planted areas.

Plant List

Symbol	Botanical Name	Common Name
A	*Achillea millefolium*	common yarrow
B	*Aquilegia formosa* var. *truncata*	red columbine
c	*Salix nivalis*	alpine willow
D	*Arctostaphylos patula*	greenleaf manzanita
E	*Populus balsamifera* ssp. *trichocarpa*	black cottonwood
F	*Carex amplifolia*	blue-leafed sedge
G	*C. tumulicola*	Berkeley sedge
H	*Eriophorum gracile*	cotton grass
i	*Ceanothus integerrimus*	deerbrush
J	*Veratrum californicum*	common corn-lily
K	*Heracleum lanatum*	cow parsnip
L	*Sidalcea oregana* ssp. *spicata*	mountain hollyhock
M	*Eriogonum umbellatum*	sulfur buckwheat

N	*Swertia radiata*	monument plant
o	*Polemonium occidentale*	Jacob's ladder
P	*Lupinus albicaulis*	fragrant lupine
Q	*L. polyphyllus*	meadow lupine
R	*Sambucus racemosa* var. *microbotrys*	mountain elderberry
s	*Aconitum columbianum*	monkshood
T	*Iris missouriensis*	mountain iris
u	*Allium validum*	swamp onion
V	*Delphinium glaucum*	tower larkspur
w	*Wyethia mollis*	mountain mule's ear
X	*Castilleja applegatei*	rock paintbrush
y	*C. miniata*	meadow paintbrush
z	*Eschscholzia californica*	California poppy
not shown	*Ipomopsis aggregata*	scarlet gilia
not shown	*Juncus ensifolius*	iris-leafed rush
not shown	*Mertensia ciliata*	mountain bluebells
not shown	*Mimulus cardinalis*	scarlet monkeyflower

Concept garden key.

not shown	*M. guttatus*	golden monkeyflower
not shown	*Penstemon azureus*	azure penstemon
not shown	*Senecio triangularis*	arrowhead butterwort

Trees

	Abies concolor	white fir
	A. magnifica	red fir
	Calocedrus decurrens	incense-cedar
	Pinus contorta ssp. *murrayana*	lodgepole pine

Plants to Use

Grasslike Plants

Note: Propagate all grass and sedge species from divisions or seed; be sure to remove old leaves and flower stalks once a year in the fall.

Sedges (*Carex* spp.).
The many sedges range from turf-forming miniatures that stand no more than three inches high to ones that form clumps two or more feet tall. Sedges have narrow, channeled, grasslike leaves arranged in three rows, and stems that carry spikes of male and

female, wind-pollinated flowers above their leaves early in the season. Male flowers have pale yellow stamens; female, feathery whitish stigmas. Avoid strongly rhizomatous species with the potential for becoming invasive. Berkeley sedge (*C. tumulicola*) is a clumping, evergreen, grasslike perennial to 15 inches high featuring dark green, recurved leaves and insignificant spikes of flowers. It accepts many soils, full sun to light shade, and occasional water. Propagate by division of clumps. Or try the low-growing *C. pansa*, which is densely short-rhizomed. Also sometimes available is the taller, blue-green-leafed *C. amplifolia*. There are many additional sedges, but they are difficult to find in the trade. Garden Design Note: *Carex pansa* can be used as a turf substitute. Needs summer water. Add taller rushes and mountain iris for accents.

Berkeley sedge (*Carex tumulicola*).

Mountain muhly (*Muhlenbergia montana*).
Densely tufted, spreading grass to 15 inches high with narrow blades. Open panicles of spikelets. Prefers dry edges of meadows.

Mountain oatgrass (*Danthonia intermedia*).
Low, turf-forming grass a few inches high. Flowering stalks to eight inches high with dense spike-lets. Tolerates a range of soils, from dryish to fairly wet.

Bent grass (*Agrostis scabra*).
Medium-sized bunch-grass—sometimes rhi-zomatous—to a foot high, taller in flower. Airy, open panicles of tiny reddish florets in summer create a red-dish haze. May be-come invasive. Adapted to wet conditions.

June grass (*Koeleria macrantha*).
Small bunchgrass eight to 12 inches high with interrupted spikes of spikelets in mid-summer. Adaptable, but best in drier areas of meadows.

Tufted hair grass (*Deschampsia caespitosa*).
Medium-sized bunchgrass to two feet high in bloom, with moderately coarse, green leaves and graceful, feathery panicles of red-tinted flowers. Cut back old leaves at year's end and recycle to the compost pile.
Garden Design Note: Subspecies *holciformis* and 'Jughandle' are lower and wider forms with attractive flowers. Excellent for erosion control. Likes moist, shady areas.

TOP: Clumps of mountain muhly (*Muhlenbergia montana*). • BOTTOM: Flowering clumps of mountain bent grass (*Agrostis scabra*).

Needle-and-thread grass (*Hesperostipa comata*).
Medium-sized bunchgrass to three feet high in bloom, with moderately coarse, pale green leaves and dramatic, open panicles of slightly nodding flowers with very long, needlelike awns. Remove old leaves from clumps at end of year. Best on the dry edge of a meadow with good drainage.

Tall Perennial Wildflowers

Note: Most perennials should have their old stems and leaves removed in fall before their winter's rest; they make a good mulch or can be used in a compost pile.

California coneflower (*Rudbeckia californica*).
Six feet tall with ovate, irregularly slashed leaves and spectacular flower heads of golden rays surrounding long cones of dark purple disc flowers. Blooms mid to late summer. Attractive to many pollinators. Propagate from divisions or seed.
Garden Design Note: Dies back early when water is withheld. Difficult to find.

Tower larkspur (*Delphinium glaucum*).
Grows from a woody rootstock to six feet tall with broad, palmately lobed, delphinium-

Flowering plants of California coneflower (*Rudbeckia californica*).

like leaves and narrow spikes of deep blue, spurred flowers. Attractive to butterflies and bumblebees; leaves poisonous. Propagate from seed.

Garden Design Note: Spectacular massed along a stream bank. If only this species were readily available.

Monkshood (*Aconitum columbianum*).
Similar overall to tower larkspur but the flowers are hooded instead of spurred. Flower color ranges from white to blue-purple. Attractive to bumblebees but poisonous to mammals. Propagate from seed.

Garden Design Note: Find it growing along mountain streams; difficult to find in nurseries.

Cow parsnip (*Heracleum lanatum*).
Bold, ragged, deeply pinnately slashed leaves with an unusual odor. Giant, hollow flower stalks to eight feet tall with broad, compound umbels of white flowers. Attractive to many pollinators. Winged fruits follow. Propagate from seed.

Garden Design Note: Takes a few years to become established. Grows best in moist shaded areas. Attracts Lorquin's admiral butterfly.

Brewer's angelica (*Angelica breweri*).
Similar overall to cow parsnip, but leaves are divided into several narrowly ellipti-

cal segments . Odor is reminiscent of celery. Flower umbels almost globe shaped; flowers white. Propagate from seed.

Showy milkweed (*Asclepias speciosa*).

LEFT: Flowering plant of cow parsnip (*Heracleum lanatum*). •RIGHT: Leaves and flowers of showy milkweed (*Asclepias speciosa*).

From wandering roots; leafy stalks to three feet high with umbels of pale pinkish purple, curiously constructed, fragrant flowers. Attractive to many pollinators. Poisonous milky sap in stems and leaves. Food plant for monarch butterfly larvae. Blooms early to mid-summer. Propagate from root divisions or seed. May be invasive.

Garden Design Note: Every butterfly garden deserves showy and fragrant milkweed. Likes dry gravelly areas. Plant with companions that allow it to ramble, or contain it with root barriers.

Shorter Perennial Wildflowers

Yampah (*Perideridia* spp.).
Grows from clusters of edible, tuberous roots. Inconspicuous leaves and small, lacy, compound umbels of white flowers on stalks to 18 inches high. Blooms midsummer. Propagate from tubers or seed.

Montane meadow plant community.

CLOCKWISE FROM TOP: Mountain elderberry (*Sambucus racemosa* var. *microbotrys*) with granite outcroppings. • Jacob's ladder (*Polemonium occidentale*).• Close-up of swamp onion (*Allium validum*). • Mountain meadow detail of mule's ears (*Wyethia mollis*) and meadow paintbrush (*Castilleja miniata*).

Garden Design Note: Beautiful massed. Unfortunately, it is not available in the trade.

Sneezeweed (*Helenium bigelovii*).
Leafy stalks from 18 inches to two feet high carry large, yellow daisies. Each head has several bright yellow, drooping rays and a buttonlike center of dark purple disc flowers. Attractive to many pollinators; blooms midsummer. Likes wet places. Propagate from divisions or seed.

Mountain hollyhock and mountain checkerbloom (*Sidalcea oregana* var. *spicata* and *S. glaucescens*).
Similar in overall appearance to common checkerbloom (*S. malviflora*), with basal tufts of round, palmately lobed leaves on short rhizomes and narrow racemes of pale purple (*S. glaucescens*) or pink (*S. oregana*) hollyhocklike flowers in midsummer. Propagate from divisions or seed.
Garden Design Note: Likes moisture.

Meadow potentilla (*Potentilla gracilis*).
Basal clumps of palmately compound, coarsely serrated leaves and flowers on stalks to a foot high with a succession of bright yellow, roselike flowers in summer. Propagate from divisions or seed. Fast grower.
Garden Design Note: Needs summer water.

Jeffrey's shooting stars (*Dodecatheon jeffreyi*).
Early blooming with many basal, spatula-shaped leaves. Naked flowering stalks to two feet high with umbels of rose- or pink-purple, cyclamenlike flowers of great beauty. Propagate from divisions or seed. Likes wet conditions.
Garden Design Note: Largest of the native shooting stars. Incredible beauty when massed in a wet meadow. Difficult to find in the trade.

Horsemint (*Agastache urticifolia*).
Clumped perennial with scented, toothed, ovate leaves and narrow spikes of two-lipped, pale purple, tubular flowers in midsummer. Flower buds deep purple. Attractive to bumblebees and butterflies. Propagate from divisions or seed.
Garden Design Note: This is a striking plant with lovely flowers, but unfortunately it is not commonly available.

Blue mountain flax (*Linum lewisii*).
Winter-dormant perennial to three feet high in flower. Wandlike stems with narrow leaves; single, flat, sky-blue flowers from late spring to summer. Requires well-drained

soil, full sun, occasional water. Propagate from seed. Often short-lived in gardens.

Garden Design Note: 'Appar' has outstanding vigor and a long season of bloom.

Mountain bluebells (*Mertensia ciliata*).
Broad clumps of oval, bluish green leaves from rhizomes. Flowering stalks rise above leaves with coiled clusters of nodding, bell-shaped, blue flowers. Flower buds pink. Propagate from divisions or seed. Tolerates wet conditions.

Garden Design Note: Not readily available.

Meadow paintbrush (*Castilleja miniata*).
Bunches of upright stems to three feet high with lance-shaped leaves and dense spikes of red, deep pink, or orange flowers. Spikes look as though they've been dipped in paint. Attractive to hummingbirds. Roots are weakly parasitic on small shrubs and grasses; sow seeds among appropriate host plants for best results.

Garden Design Note: Not readily available; difficult to establish.

Mountain iris (*Iris missouriensis*).
Stiff, sword-shaped, equitant leaves from rhizomes. Elegant, flaglike, pale blue flowers, with falls prettily striped with dark purple and yellow. Attractive to many pollinators; flowers early summer. Clumps increase to sizable circles when conditions are

TOP: Flowers of blue mountain flax (*Linum lewisii*). • BOTTOM: Mosaic of lupines, sagebrush (*Artemisia* sp.), and meadow paintbrush (*Castilleja miniata*).

right. Adapted to somewhat alkaline soils.

Garden Design Note: Difficult to grow away from its home-land. Propagate from divisions or seed.

Jacob's ladder (*Polemonium occidentale*).

Clumps of pinnately com-pound, pealike leaves and flowering stalks to a foot tall with a long succession of sky

blue, saucer-shaped flowers enhanced by bright yellow stamens. Propagate from seed; may self-sow.

Garden Design Note: Outstanding wet meadow favor-ite. Should be available but is not.

Ground Covers

Note: These tiny perennials seldom leave enough old leaves to interfere with new growth and so are basically maintenance free.

Primrose monkeyflower (*Mimulus primuloides*).

Miniature clumps of narrowly elliptical leaves covered with slender white hairs; thread-like stems bear a single, clear yellow, miniature monkeyflower, blooming midsummer. Prefers moist areas. Propagate from divisions.

Garden Design Note: A ground cover for moist conditions.

Blue violet (*Viola adunca*).

Slowly creeping perennial with nearly round, dark green leaves and short stalks that bear a single pale to deep blue, violet-shaped flower. Choose clones with deep colors. Blooms sporadically from spring into summer. Propagate from divisions or seed. May become invasive.

Garden Design Note: Not commonly available.

White meadow violet (*Viola macloskeyi*).

Miniature creeping violet with round leaves and short flower stalks with curiously shaped, snowy white flowers marked with purple lines. Prefers wet places; blooms

TOP: Mountain iris (*Iris missouriensis*). • BOTTOM: Jacob's ladder (*Polemonium californicum*).

early spring. Propagate from divisions. Garden Design Note: Not easily found in nurseries.

Tinker's penny (*Hypericum anagalloides*).
Prostrate, creeping, herbaceous ground cover with tiny oval leaves and miniature, salmon yellow flowers sprinkled over the mats, blooming spring into summer. Prefers wet places. Propagate from divisions.

Bulbs

Swamp onion (*Allium validum*).
Bold, wet-grower from rhizomatous bulbs. Upright leaves to two feet high and stems to three feet, with dense umbels of pale to deep pink, vase-shaped flowers. Blooms midsummer. Propagate from bulb divisions.

Garden Design Note: Mass these along a stream. Swamp onion must not be allowed to dry out.

Pretty face brodiaea (*Triteleia ixioides* var. *anilina*).

Low-growing corm with narrow, strap-shaped leaves and stems a few inches high with open umbels of pale yellow, starlike flowers, each petal with a central dark purple stripe. Blooms midsummer. Propagate from offsets or seed.

Camas (*Camassia quamash*).
Vigorous, folded, glossy green leaves and dense racemes of starry, clear to deep blue flowers that open just after snowmelt. Prefers wet places. Propagate from offsets.
Garden Design Note: This large and showy wet-meadow bulb is readily available, with many cultivars and selections. Bulbs are edible.

TOP: Swamp onion (*Allium validum*, rose-pink flowers) with meadow paintbrush (*Castilleja miniata*). • BOTTOM: Flower spike of camas (*Camassia quamash*).

Alpine lily (*Lilium parvum*).

Scaly bulbs bear stems to four feet high with spiraled leaves and several small, horizontally held, bell-shaped, dark orange flowers, the petals sprinkled with dark spots. Prefers wet places. Propagate from bulb scales or seed.

Garden Design Note: When I see this lily in bloom, I always take a closer look. Its many bell-shaped flowers, smaller than Humboldt or leopard lilies, are charming.

Kelley's lily (*Lilium kelleyanum*).

Scaly bulbs bear stems to six feet high with whorls of leaves and several nodding, fragrant, orange flowers decorated with purple spots. Striking; similar to leopard lily. Prefers wet places. Propagate from bulb scales or seed.

Additional grasses to try:

Squirreltail (*Elymus elymoides*), DRY MEADOWS; flag melic (*Melica stricta*), ROCKY MEADOWS; Brewer's blue grass (*Poa breweri*); deer grass (*Muhlenbergia rigens*).

Additional tall perennials to try:

Corn-lily (*Veratrum californicum*), ranger's buttons (*Sphenosciadium capitellatum*), green gentian or monument plant (*Swertia radiata*), fireweed (*Epilobium angustifolium*), meadow knotweed (*Polygonum phytolaccaefolium*), mountain elderberry (*Sambucus racemosa* var. *microbotrys*).

Additional short perennials to try:

Mountain lovage (*Ligusticum grayi*), pink plumes (*Geum trifidum*), Brewer's cinquefoil (*Potentilla drummondii* ssp. *breweri*), meadow buttercup (*Ranunculus alismaefolius*), whorled penstemons (*Penstemon rydbergii* and *P. heterodoxus*), mountain pride (*P. newberryi*), red columbine (*Aquilegia formosa*), pink monkeyflower (*Mimulus lewisii*), golden monkeyflower (*M. tilingii*), elephant snouts (*Pedicularis groenlandica*), marsh-marigold (*Caltha leptosepala*), mountain mule's ears (*Wyethia mollis*), wanderer's gentian (*Gentiana calycosa*).

Alpine lily (*Lilium parvum*).

Places to Visit

Mt. Eddy, Trinity Mountains, Siskiyou County. Mt. Eddy is a sugarloaf-shaped mountain directly west of the towns of Weed and Mt. Shasta. The trailhead is accessed from the top shoulder of Stewarts Springs Road, located west of Hwy 5 at the north end of Weed. The Pacific Crest Trail takes you three miles through mixed conifer forest, hanging meadows, and a bog-meadow ecosystem (here the trail splits, and a left-hand spur gives access to the summit of the mountain). Mid-July is the best time to visit for meadow wildflowers. Highlights include cobra plant (*Darlingtonia californica*), white meadow violet (*Viola macloskeyi*), common corn-lily (*Veratrum californicum*), tower larkspur (*Delphinium glaucum*), monkshood (*Aconitum columbianum*), golden monkeyflowers (*Mimulus guttatus* and *M. tilingii*), snowy rein orchid (*Platanthera dilatata*), alpine shooting star (*Dodecatheon alpinum*), red columbine (*Aquilegia formosa*), mountain-marigold (*Caltha leptosepala*), butterworts (*Senecio* spp.), Brewer's angelica (*Angelica breweri*), cow parsnip (*Heracleum lanatum*), meadow star-tulip (*Calochortus nudus*), and leopard lily (*Lilium pardalinum*).

Winnemucca Lake, central Sierra, Alpine County. Take Hwy 88 to the top of Carson Pass. A large parking lot there marks the trailhead for an easy two-and-a-half-mile hike to Winnemucca Lake. The trail passes through fine subalpine forest and extensive dry and hanging meadows before arriving at the lake. The area boasts great species diversity due to the variety of soils and an abundance of water that flows from snowmelt. Behind the lake is alpine scree with Sierra primrose (*Primula suffrutescens*), alpine buttercup (*Ranunculus eschscholzii*), white-heather (*Cassiope mertensiana*), bog-laurel (*Kalmia polifolia* var. *microphylla*), and red-heather (*Phyllodoce breweri*). Meadows include marsh marigold, tower larkspur, horsemint (*Agastache urticifolia*), red columbine, mountain iris (*Iris missouriensis*), mountain bluebells (*Mertensia ciliata*), elephant snouts (*Pedicularis groen-landica*), baby elephant snouts (*P. attollens*), alpine shooting star, meadow paintbrush (*Castilleja miniata*), mountain lovage (*Ligusticum grayi*), and gentians (*Gentiana* spp.).

Parcher's resort, Bishop Creek, east side of the southern Sierra, Inyo County. Take Hwy 168 west from Bishop. Stay left at the fork and drive to the end of the road at South Lake. A trail takes you approximately a mile down to Parcher's resort. The resort and the trailhead both have excellent high-mountain meadow habitat. Among the lodgepole pines (*Pinus contorta* ssp. *murrayana*), western white pines (*P. monticola*), mountain alders (*Alnus incana* var. *tenuifolia*), and quaking aspens (*Populus tremuloides*), the meadow plants include Kelley's lily (*Lilium kelleyanum*), monkshood, tower larkspur, water hemlock (*Cicuta douglasii*), dotted saxifrage (*Saxifraga odontoloma*), grass-of-Parnassus (*Parnassia palustris*), ranger's buttons (*Sphenosciadium capitellatum*), monkeyflowers (*Mimulus* spp.), fireweed (*Epilobium angustifolium*), gentians, meadow butterwort (*Senecio triangularis*), snowy rein orchid, green rein orchid (*Platanthera sparsiflora*), and red columbine.

Crane Flat, Yosemite National Park, Tuolumne County. Crane Flat is located where the Tioga Pass and Big Oak Flat roads meet. Near the service station and store, meadows wind their way for over a mile amid mixed conifer forest. The meadow flowers include cow parsnip, horsemint, alpine lily (*Lilium parvum*), common corn-lily, meadow St. John's wort (*Hypericum formosum*), snowy rein orchid, yampah (*Perideridia* sp.), camas (*Camassia quamash*), Jeffrey's shooting star (*Dodeca-*

theon jeffreyi), green gentian (*Swertia radiata*), pretty face brodiaea (*Triteleia ixioides* var. *anilina*), mountain lovage, meadow lotus (*Lotus oblongifolius*), sneezeweed (*Helenium bigelovii*), and California coneflower (*Rudbeckia californica*).

Note: Although the high mountains of southern California—especially the San Bernardinos, San Gabriels, and San Jacintos—have montane meadows, they are generally less extensive and diverse than the meadows in the Sierra Nevada, Cascade Mountains, and Klamath Ranges.

Cow parsnip (*Heracleum lanatum*) and corn-lily (*Veratrum californicum*) meadow, Crane Flat, Yosemite National Park.

CHAPTER 7

MIXED-EVERGREEN FOREST: SUMMER SHADE BETWEEN FOG AND SUN

Between the open oak woodlands of the hot foothills and the foggy coastal redwood forests lies a wooded terrain that combines aspects of both these environments. Mixed-evergreen forest is a descriptive name: a densely wooded habitat of various, usually evergreen, trees adapted to summers of intermittent fog and clear hot days, and to moderate winter rainfall. The trees here have extensive thirsty roots, and their canopy creates a deep, dry summer shade; they thus challenge understory plants with a combination of low water and little light.

Mixed-evergreen forests cover much territory in California's coastal mountains. The farther inland you go, the more this forest gives way to oak woodland, and its last stands cling to canyon bottoms and steep, north-facing slopes. In southern California, this community is limited to deep canyons due to increasingly arid conditions, but in central and northern California this forest covers broad swaths of land. Many sites around San Francisco Bay support mixed-evergreen forest, including the head of the Napa Valley, Annadel State Park in Sonoma County, Mt. Tamalpais in Marin County, and much of the

peninsula south of San Francisco. In the Sierra Nevada, a modified version of this forest occurs along the lower fringes of the yellow pine belt, where ponderosa pine (*Pinus ponderosa*), sugar pine (*P. lambertiana*), incense-cedar (*Calocedrus decurrens*), and white fir (*Abies concolor*) often mix with canyon live oak (*Quercus chrysolepis*), black oak (*Q. kelloggii*), California bay (*Umbellularia californica*), madrone (*Arbutus menziesii*), big-leaf maple (*Acer macrophyllum*), Douglas-fir (*Pseudotsuga menziesii*), California nutmeg (*Torreya californica*), and others.

LEFT: Dense forest of tanbark oak (*Lithocarpus densiflorus*) and madrone (*Arbutus menziesii*). • ABOVE: Stand of mixed evergreens on a north-facing slope in the Santa Lucia Mountains.

Different areas of mixed-evergreen forest experience varying degrees of summer fog and winter rain, resulting in a composition that changes from site to site. For example, cool coastal conditions favor Douglas-fir, tanbark oak (*Lithocarpus densiflorus*), and coast chinquapin (*Chrysolepis chrysophylla*), while warmer, sunnier conditions support canyon live oak and coast live oak (*Quercus agrifolia*), but no conifers. Conditions in between support madrone, California bay, and California nutmeg, which also often appear in coastal versions of this forest. An-

TOP: Pink-flowering currant (*Ribes sanguineum* var. *glutinosum*) next to creek. • BOTTOM: Serviceberry flowers (*Amelanchier* sp.).

170

other factor affecting forest composition is fire. Frequent fire favors the faster-growing, shorter broadleaf trees such as oaks and madrone. In the absence of fire, Douglas-firs tower above these trees, eventually shading them out. When gardening in this community, it's important to know which face this forest is likely to wear in your particular area.

Just as the main canopy varies from place to place, so too does the understory. Various shrubs thrive in the better-lighted areas, along with a number of drought-tolerant perennials, ferns, grasses, and bulbs. On gentle slopes where summer fog adds moisture, the understory approaches the lushness of redwood forests; on steep slopes with only occasional fog, the understory is patchy and sparse, presenting a dusty, dry appearance in summer.

Creating a Garden

By coming to know individual sites in this varied community, you'll soon learn whether the understory plants are summer drought-tolerant or appreciate summer water. The suggested plant palette at the end of the chapter covers both categories.

Gardening in this community offers rewards and challenges: the trees are among

Woodland garden with lady ferns (*Athyrium filix-femina*) and alumroot (*Heuchera* 'Wendy'). Photograph by S. Holt.

CLOCKWISE FROM TOP: Cottage style garden with chaparral and grassland elements. • Entryway to cottage with raised brick planters. • Chaparral and grassland plants in a mixed-evergreen woodland garden.

California's most beautiful, but finding the right understory plantings may take some effort. The rule is to err on the side of simplicity, with just enough elements of surprise to enliven the scene—using patches of colorful spring perennials and bulbs, for example. Harmonizing elements—ground covers, shade-tolerant grasses, and drought-tolerant ferns—can be very useful in this setting.

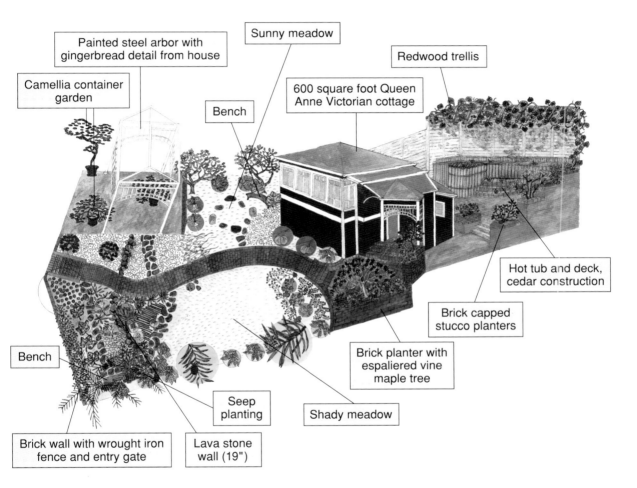

Design Notes

This garden was designed to complement a charming Victorian cottage built circa 1910, situated near the rear of a long, narrow lot. The house has been beautifully restored and is part of a quaint historic neighborhood in southern San Francisco. Though this hilly area receives some summer fog, it is known by San Franciscans for its sunny days.

Rendering of featured garden.

Existing on site were two coast redwoods, a Hollywood juniper, Monterey pine, deodar cedar, and European birch. Many other plants were removed and adopted into other gardens. The long-term plan includes planting native trees (Santa Lucia fir and California nutmeg) that will eventually replace the birch and cedar. With the additional exception of a *Heuchera sanguinea* (native to Arizona and New Mexico), all other plants in the new garden were selected from native plant communities commonly found in California, with mixed-evergreen forest, redwood and riparian forests (seep), chaparral, a shady wet meadow, and a dry sunny meadow being most prominently featured.

A cottage theme seemed ideal for this garden, with a series of paths and walkways to connect the various woodland settings. Because the grade slopes upward toward the house, a terrace helped make the redwood garden more user-friendly. A dark gray volcanic rock wall provides visual interest, while a volcanic rock stepping-stone path leads to a bench beneath the redwoods, and then passes by a wet and low area at the edge of the shady wet meadow. Opposite the redwood garden is the roof of the garage, which the client uses as a patio. A welded and painted steel trellis was proposed but not built; repeating the wood trim details over the porch entry, it would have given this rooftop patio a distinctive presence. A collection of camellias in oversized red Chinese pots completes the effect.

The dry meadow also has a stepping-stone path, which leads to another bench.

Clockwise from top: California nutmeg (*Torreya californica*) • Firecracker brodiaea (*Dichelostemma ida-maia*) • Paper onion (*Allium unifolium*) • Yerba buena (*Satureja douglasii*) • Flowering currant (*Ribes sanguineum*).

This path includes four large flat stones, in red, black, green, and white, selected by the owner to represent the four cardinal directions in Native American culture. The meadow becomes ablaze with wildflower color in the spring. It also is home to a number of rare California native bulbs that we were fortunate enough to obtain in 1999.

As one approaches the entry on the curved brick walkway, a linear planter becomes visible against the property-line fence. An espaliered vine maple, underplanted with island alumroot, gives bold definition to this space. A second entry planter gets full afternoon sun. Here a bigberry manzanita is underplanted with one of our showiest natives, lilac verbena, a profuse bloomer that enlivens the garden.

The rear garden includes mostly chaparral species in raised brick planters. Golden currant is espaliered to one garden wall, while twinberry and California grape ramble over a wooden trellis constructed above a cedar soaking tub. A bonsai shore pine to the left of the stairs completes the setting.

Scope of Work

- Demolish existing garden, recycling unwanted plants to friends and neighbors.
- Grade flat area adjacent to redwoods.
- Select lava rocks and flat stones of different colors, with owner's participation.
- Create paths and add stepping stones and boulders.
- Build lava rock wall.
- Weld steel gazebo, attach it to the garage roof, and paint it to match the house trim.
- Install a combination of drip and overhead spray irrigation. Install extra irrigation in the seep area. Install drip lines to containers.
- Seed meadows.
- Install plants.
- Install mulch.
- Add planters and decorative pots.
- Install low-voltage lighting in existing trees.

Golden currant (*Ribes aureum*) espaliered against a wall with twinberry honeysuckle (*Lonicera involucrata*) and California grape (*Vitis californica*) climbing on a trellis (featured garden).

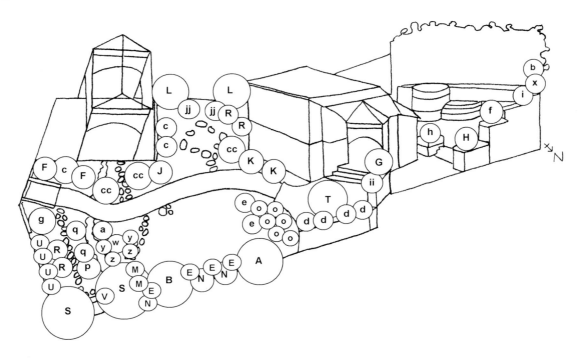

Plant List

SYMBOL	BOTANICAL NAME	COMMON NAME
A	*Abies bracteata*	Santa Lucia fir
	Allium unifolium	paper onion
B	*Torreya californica*	California nutmeg
	Dichelostemma ida-maia	firecracker brodiaea
C	*Styrax officinalis* var. *redivivus*	snowdrop bush
D	*Thalictrum fendleri* var. *polycarpum*	foothill meadow-rue
E	*Satureja douglasii*	yerba buena

Chaparral Community

F	*Ceanothus griseus* var. *horizontalis* 'Yankee Point'	California lilac
G	*Arctostaphylos manzanita*	common manzanita
H	*A. pajaroensis*	Pajaro manzanita
i	*A.* 'Sunset'	Sunset manzanita
J	*Salvia sonomensis*	Sonoma sage

Featured garden key.

Oak Woodland Community

| K | *Cercis occidentalis* | western redbud |
| L | *Heteromeles arbutifolia* | toyon |

Redwood Community

M	*Aquilegia formosa*	red columbine
N	*Myrica californica*	California wax myrtle
O	*M. californica* 'Buxifolia'	California wax myrtle
P	*Oxalis oregana*	redwood sorrel
q	*Rhododendron occidentale*	western azalea
R	*Ribes sanguineum* var. *glutinosum*	pink-flowering currant
S	*Sequoia sempervirens*	coast redwood
T	*Acer circinatum*	vine maple
U	*Vaccinium ovatum*	evergreen huckleberry
V	*Asarum caudatum*	wild ginger

Riparian Community

W	*Juncus patens*	California rush
X	*Lonicera involucrata*	twinberry honeysuckle
Y	*Lilium pardalinum*	leopard lily
Z	*Mimulus guttatus*	golden monkeyflower
a	*Calycanthus occidentalis*	western spice bush
b	*Vitis californica* 'Roger's Red'	California grape

Coastal Bluff Community

| cc | *Baccharis pilularis* 'Pigeon Point' | dwarf coyote brush |

Other Plant Communities: Channel Islands, Closed-Cone Forest, and South Coastal Scrub

d	*Heuchera maxima*	island alumroot
e*	*H. sanguineum*	coral bells
f	*Ribes aureum*	golden currant
g	*R. viburnifolium*	evergreen currant
h	*Pinus contorta* ssp. *contorta*	shore pine (bonsai)

* In Arizona and New Mexico, substitute *Heuchera* 'Wendy' or *H.* 'Santa Ana Cardinal'.

| ii | *Verbena lilacina* | lilac verbena |
| jj | *Berberis pinnata* 'Golden Showers' | California barberry |

Plants to Use

Trees

Canyon live oak (*Quercus chrysolepis*).
Multitrunked tree to 60 feet tall with massive branches and pale whitish bark. Handsome, lance-shaped leaves are glossy green above, golden to grayish below, and smooth-margined or lined with prickly teeth. Drought tolerant. Propagate from fresh acorns. Garden Design Note: Large trees grown from acorns are available from specialty tree nurseries. Moderate growth rate and adaptable to a variety of soils.

Douglas-fir (*Pseudotsuga menziesii*).
Conifer to over 200 feet tall in old age. Massive trunk with irregularly fissured bark. Lemon-scented needles spirally arranged on branches. Distinctive hanging seed cones with rounded scales and three-pronged bracts. Grows taller than other trees of this

Mixed evergreen woodland with a gravel path. Photograph by S. Morris.

forest but only at a moderate rate (fast in its first few years). Appreciates some summer water. Propagate from seed. In southern California, use big-cone Douglas-fir (*P. macrocarpa*).

Garden Design Note: Smaller cultivars for urban gardeners are available at dwarf conifer nurseries in Oregon. The fresh new growth is high in vitamin C and makes a refreshing tea.

California nutmeg (*Torreya californica*).

Small conifer to 40 or 50 feet tall. Stump sprouts, so can be used as a hedge. Long, glossy needles are spine-tipped. Male trees produce tiny cream-colored pollen cones; female trees make plumlike seed cones. Moderately drought tolerant. Propagate from cuttings or seed.

Garden Design Note: In hot areas, this tree likes a cool, north-facing slope. Needles are very sharp.

Madrone (*Arbutus menziesii*).

Large tree to 70 feet tall, often multi-trunked with peeling brown old bark and smooth orange-brown new bark. Large, handsome, broadly elliptical leaves with finely serrated margins, deep green above and silvery beneath. Dense panicles of fragrant, white, urn-shaped flowers in early to midspring. Orange-red, warty berries in fall. Occasional summer water when established. Propagate from stratified seed. Often difficult to establish; mycorrhizal innocula should improve success rate. Garden Design Note: To establish one of these trees, I plant three. Tremendous wildlife value: both open-nesting and cavity-nesting birds favor the madrone, and butterflies visit its flowers.

California bay (*Umbellularia californica*).

Large, often multitrunked tree to 70 feet tall with scaly, dark brown to smooth bark. Glossy, dark green, highly fragrant, lance-shaped leaves. Small clusters of saucer-shaped, pale yellow flowers from winter to early spring. Drupelike purple, avocado-shaped fruits

Sculpted trunk of a young madrone (*Arbutus menziesii*) in a garden. Photograph by S. Ingram.

in fall. Occasional summer water best. Propagate from seed or stump sprouts. Stump sprouts should be regularly pruned out unless a grove of bays is eventually planned for. Can also be pruned into a screen or hedge.

Garden Design Note: Slow-growing and allelopathic. Stellar jays, Clark's nutcrackers, Townsend's solitaires, grosbeaks, squirrels, and the California mouse eat the seeds. Leaves can be used in cooking. Easy to hold at six to eight feet in height. Can take shade. Growth will be more rapid in rich soil with cool, moist growing conditions.

Shrubs

Snowdrop or snowbell bush (*Styrax officinalis* var. *redivivus*).
Deciduous shrub to 12 feet high with arching branches. Round leaves; fragrant white bell-shaped flowers hang under branches in mid to late spring. Requires well-drained

Peeling bark of a madrone (*Arbutus menziesii*).

soil, full sun to light shade, little water. Propagate from fresh seed or suckers. Prune to shape, but don't over-do it; snowdrop bush grows more slowly than many other shrubs, so excessive pruning takes a long time to correct. Garden Design Note: Punctuate a sunny slope with snowdrop bush, then fill in with faster-growing ground covers.

Pink-flowering currant (*Ribes sanguineum* var. *glutinosum*).

Fast-growing shrub to 12 feet tall with stiff, smooth branches. Deciduous, fragrant, sticky, palmately lobed leaves; trusses of pink flow-ers in late winter to early spring (snowy white forms available as well); pale purple berries. Appreciates occasional summer water. Propa-gate from cuttings in winter or early spring. Pink-flowering currant can be cut partway back to maintain a denser structure. Shrubs may suddenly die from fungal infection, so re-starting the shrubs from cuttings every few years is a good idea.

Garden Design Note: Among the many unique selections, I especially like 'King Ed-ward VII'. It blooms early, with flowers that are deep pink, pendulous droplets. Its foliage adds red-orange fall color, too. Plant flower-ing currant in thickets at the edge of the shade, or espalier singly against an east-facing wall that complements the bright flowers.

TOP: Flowering branches of snowdrop bush (*Styrax officinalis* var. *redivivus*). • BOTTOM: Pink-flowering currant (*Ribes sanguineum* var. *glutinosum*) in peak bloom.

Creambush (*Holodiscus discolor*).
Dense, fast-growing shrub to 15 feet tall with fine branches. Leaves are small, deciduous, fruit-scented, soft, and elliptical, with coarsely toothed margins. Dense panicles of tiny pink-budded, cream-colored flowers appear in late spring or early summer. Relatively drought tolerant. Propagate from suckers, cuttings, or seed. Every few years, thin the dense, inner branches to avoid an unsightly tangle.

Garden Design Note: This is a very versatile shrub that shows best when massed along a path. Plant on a rise so the arching blossoms spill toward you.

Wood rose (*Rosa gymnocarpa*).
Fast-growing, colonizing small shrub to four feet tall with prickly stems. Deciduous, pinnately compound leaves. Small, pale to dark pink, fragrant, single-rose blossoms in midspring. Small, deep orange hips in late summer or fall. Needs occasional summer water. Propagate from divisions or suckers. Wood rose canes can be cut to the ground in winter to encourage healthy new growth in spring.

Garden Design Note: You may have difficulty containing this shrub. Plant it along water courses or depressions where it can make low thickets. Excellent habitat plant.

Deerbrush (*Ceanothus integerrimus*).
Fast-growing shrub to 12 feet tall with divergent, sometimes thorny branches. Thin, deciduous, elliptical leaves with finely toothed margins. Long, dense panicles of frothy white, lavender, pale pink, or blue, fragrant flowers in late spring.

TOP: Flowering branches of creambush (*Holodiscus discolor*). • BOTTOM: Flowering sprays of deerbrush (*Ceanothus integerrimus*).

182

Needs occasional summer water. Propagate from hardwood cuttings in fall or stratified seed. Prune out dead branches.

Garden Design Note: Although this species is not commonly grown, it is a useful one. I like 'Ray Hartman', 'Sierra Blue', and 'Snowball', all of which require pruning and shaping for optimum garden use. Create berms, plant on slopes, plant with boulders or in crevices, and withhold excessive summer water. If managed properly, these hybrids and selections will live for 25 years.

Lemmon's ceanothus (*Ceanothus lemmonii*).

Low-growing, semisprawling shrub to around three feet tall. Tough, evergreen, broadly elliptical, dark green leaves. Dense panicles of intensely blue, fragrant flowers in midspring. Drought tolerant. Propagate from hardwood cuttings in fall or stratified seed.

Garden Design Note: Another species that is not commonly grown. For semisprawling California lilacs that will perform well in mixed-evergreen woodlands, plant 'Yankee Point', 'Joyce Coulter', and 'Arroyo de la Cruz'.

Azalea-flowered monkeyflower (*Mimulus bifidus* [= *M. aurantiacus* in *The Jepson Manual*]).

Small, fast-growing shrub to around two and a half feet high. Pairs of tough, shiny, evergreen, narrowly lance-shaped leaves. Large, two-lipped, pale apricot to near white flowers with orange throats grow in small clusters and resemble azaleas from a distance. (Many hybrid cultivars of bush monkeyflowers feature this species as a parent.) Requires occasional summer water and good drainage. Propagate from tip cuttings or seed. It's a good idea to start new

Flowering shrub of azalea-flowered monkeyflower (*Mimulus bifidus*).

plants from cuttings every couple of years, as old plants get leggy and die back. Watch out for brittle branches that break easily.

Garden Design Note: This is my favorite monkeyflower, for both its foliage and its flowers. Great for a successional garden, as it grows fast and blooms profusely. Look for 'Salmon Creek' or 'Bowman Lake', a Ted Kipping introduction. Prefers dry, rocky places.

Grasses

California melic (*Melica californica*).

Small bunchgrass to around a foot tall, with many upright, bright green leaves and narrow, spikelike clusters of membranous spikelets. Occasional summer water best. Propagate from divisions or seed.

Garden Design Note: I combine this with purple needlegrass and California fescue for larger hillside meadows in northern California, or at the edges of mixed woodlands.

Perennials

Starflower (*Trientalis latifolia*).

Loosely colonizing, fall/winter-dormant plant from tiny tubers. The main stems are a few inches tall and bear a whorl of several broad leaves; the small, star-shaped, white or pink flowers appear above leaves in mid to late spring on their own threadlike stems. Needs occasional summer water until plants die back. Propagate from tuberous offsets.

Siskiyou iris (*Iris innominata*).

Miniature, semievergreen, rhizomatous perennial to eight inches high in flower. Shiny, narrow, swordlike leaves in a flattened fan, with showy iris flowers in bright yellow or purple emerging in late spring. Requires well-drained soil, light shade, some water. Propagate from divisions of rhizomes or from seed. Divide clumps every three or four years so that rhizomes and roots don't become tangled and interfere with each other. In southern California, use Douglas iris (*I. douglasiana*).

Garden Design Note: This is a great addition to the woodland garden. Use on hillsides with bulbs. It will colonize.

Indian pink (*Silene californica*).

Sprawling, fall/winter-dormant perennial from deep-seated, fleshy roots. Pale green, lance-shaped leaves in pairs along stems; large, solitary, scarlet-red flowers, reminiscent of carnations. No water after summer dieback. Propagate from seed. In southern California, use *S. laciniata*.

Garden Design Note: Sometimes I find Indian pink at native plant sales. *Silene laciniata*

is available through High Country Gardens Nursery. Provide good drainage and sun protection for longer life, and mass for blazing results.

Fireweed (*Epilobium angustifolium*). Tall, winter-dormant perennial from spreading roots to five feet high. Narrow, willowlike leaves on stems that turn red or bronze in fall. Tall racemes of pink-purple flowers. Accepts many soils, full sun to light shade, occasional water. Propagate by

TOP: *Iris douglasiana* 'Canyon Snow' in a shade garden. • BOTTOM: Flowering plant of Indian pink (*Silene californica*) with fallen flowers of snowdrop bush (*Styrax officinalis* var. *redivivus*).

root divisions or seed. Cut back stems to the ground in fall to encourage new growth in the spring. May be invasive.

Garden Design Note: Prefers cooler environments like open meadows near woodlands on east- and north-facing slopes. Available from northern nurseries. Try at higher elevations in southern California.

Foothill meadow-rue (*Thalictrum fendleri* var. *polycarpum*). Fall/winter-dormant perennial from a rootstock, to three feet high when in flower. Beautiful, highly dissected, fernlike leaves. Rather insignificant, greenish, wind-pollinated flowers in spring; male and female flowers on separate plants. Cut back in late summer. Accepts extra water with drainage. Propagate from divisions or seed.

Garden Design Note: The foliage of meadow-rue is most attractive. Because it is a tall perennial, I use it at the back of the mixed woodland perennial border. Underplant with alumroot, yerba buena, and modesty.

TOP: New shoots of foothill meadow-rue (*Thalictrum fendleri* var. *polycarpum*). • BOTTOM: Crevice alumroot (*Heuchera rubescens*).

Common alumroot (*Heuchera micrantha*).
Clumped perennial from woody rootstocks and rosettes of round, scalloped leaves, often prettily marked with purple veins. Airy panicles of tiny, bell-shaped white or pink flowers to two feet high in mid to late spring. Great massed on a steep slope or used in quantity as a coarse ground cover. Many hybrids and cultivars are available in the genus. Low maintenance. Propagate from division of clumps or seed.
Garden Design Note: Heucheras come in all sizes and color variations. Such species as *H. micrantha* and *H. maxima* (island alumroot) provide pleasing results in the garden; I also also like to mass 'Wendy' or 'Canyon Delight' under oaks or mixed with California fescue, Douglas iris, and western sword fern.

Ground Covers

Yerba buena (*Satureja douglasii*).
Low, trailing plant with pairs of oval, mint-scented leaves and tiny, white flowers borne in leaf axils. Leaves make a delightful, mint-flavored tea. Prefers coastal conditions and needs good drainage. Propagate from rooted sections.
Garden Design Note: Excellent ground cover in a woodland garden. Mix with modesty for color and texture variation. Plant it in a children's garden. Both a Spanish language and cooking lesson, this "good herb" makes a great tea with cookies.

Modesty (*Whipplea modesta*).
Trailing, woody ground cover with pairs of dull green, ovate leaves and dense clusters of tiny white flowers in spring. Good in many soils and especially on inland slopes. Propagate from rooted sections.
Garden Design Note: I've successfully planted modesty under oaks and in mixed-evergreen woodlands. It's slow to establish but warrants the patience. Plant on a bank at a shady entry with heucheras and ferns.

Ferns

Propagation Note: All ferns are propagated by divisions or spores.

Lady fern (*Athyrium filix-femina*).
Large, winter-deciduous, clumping fern with lacy, twice pinnately compound fronds to five feet high. Sori are oval-shaped with a crescent-shaped indusium. Needs year-round moisture. Cut plants back to the ground in late fall.
Garden Design Note: I mix lady fern, wood fern, and western sword fern under trees

Leaves of yerba buena (*Satureja douglasii*).

in the woodland garden. Mass each species for best effect. Allow room for colonizing and growth. Available from northern nurseries in Oregon.

California maidenhair (*Adiantum jordanii*).
Small, summer-dormant, clumping fern with wiry black stalks and lacy fronds with flag-shaped segments. Sori are along the margin of the underside of the frond, with the thickened edge forming an indusium. No summer water.
Garden Design Note: Place this graceful fern near a water feature where it gets splashes or drips during the cool months.

Bulbs

Firecracker brodiaea (*Dichelostemma ida-maia*).
Two to three long, strap-shaped, basal leaves turn brown by flowering time in late spring. Naked stalk to one foot high bears an umbel of nodding, tubular, dark red flowers tipped by recurved green petals and cream-colored appendages. No water after dieback. Propagate from cormlets or seed.
Garden Design Note: This bulb is available at specialty nurseries such as Telos Rare Bulbs or Far West Bulbs.

Washington lily (*Lilium washingtonianum*).
Winter-dormant bulb to eight feet high in flower. Whorls of glossy leaves and racemes of large, fragrant, waxy white flowers in summer. Requires well-drained soil, light shade, no water after dieback. Propagate from bulb scales or seed. Difficult to establish.
Garden Design Note: I anxiously await the day when this special flower is available in the nursery trade. Spectacular growing on steep slopes in scree near the shade of firs and pines.

TOP: Bank of California maidenhair fern (*Adiantum jordanii*) with California fescue (*Festuca californica*). • LOWER RIGHT: Flowers of firecracker brodiaea (*Dichelostemma ida-maia*). • BOTTOM LEFT: Washington lily (*Lilium washingtonianum*).

Diogenes' lantern (*Calochortus amabilis*).
Single, glossy, strap-shaped leaf appearing in early spring. In midspring, a stalk with
smaller leaves to a foot tall carries nodding, lantern-
shaped, deep yellow flowers of great beauty. No water after dieback. Propagate from bulblets or seed.
Garden Design Note: Available at specialty nurseries.
Mix with firecracker brodiaea and paper onion in meadows near woodland edges; clarkia and lupines are also
good companions.

Paper onion (*Allium unifolium*).
One or two bright green, strap-shaped leaves, fading
by midspring when flowers open. Blooms on a naked
stalk to a foot high with an umbel of starlike, bright
pink (occasionally white) flowers. Needs no summer
water. Propagate from bulblets or seed.
Garden Design Note: One of the most commonly available bulbs, and easy to grow. Order plenty and plant in
drifts with Ithuriel's spear. You may want to cage the
bulbs if gophers are present.

Annuals

Note: All annuals are propagated from seed.

Red ribbons (*Clarkia concinna*).
Branched, low-growing annual
with narrowly spoon-shaped leaves
and clusters of showy, shocking
pink, wide-open flowers of great
beauty. Each petal is deeply slashed
into three parts; red, ribbonlike sepals curl down. Water extends
bloom. May self-sow.
Garden Design Note: Mix with
wind poppies and Chinese houses.
Available at Annie's Annuals and
Larner Seeds.

TOP TO BOTTOM: Flowers of Diogenes' lantern (*Calochortus amabilis*). • Flowering plant of paper onion (*Allium unifolium*).

• Bank of red ribbons (*Clarkia concinna*).

Harlequin lupine (*Lupinus stiversii*).
Few-branched annual to a foot high with dense racemes of showy, yellow-and-pink, sweet pea-shaped flowers. Basal leaves are palmately compound with obovate leaflets. Propagate from fresh seed.

TOP: California barberry (*Berberis pinnata*) with purple sage (*Salvia leucophylla*). • BOTTOM LEFT: Flagstone path with pink-flowering currant (*Ribes sanguineum* var. *glutinosum*) and California lilac (*Ceanothus* sp.). • BOTTOM RIGHT: Serviceberry (*Amelanchier* sp.).

Garden Design Note: One of the showiest lupines. Not readily available, but easy to grow from seed, if added to boiling water then allowed to cool in the water. Plant immediately. *L. microcarpus* var. *densiflorus* 'Ed Gedling' is a good substitute.

Bride's bouquet (*Collomia grandiflora*).
Few-branched annual to two feet high with lance-shaped leaves and dense heads of tubular, pale apricot flowers complemented by blue stamens. Water extends bloom. May self-sow.
Garden Design Note: Sometimes available from Annie's Annuals. Needs sun and good drainage.

Additional shrubs to try:

Canyon gooseberry (*Ribes menziesii*), toyon (*Heteromeles arbutifolia*), California barberry (*Berberis pinnata* 'Golden Abundance'), pitcher sage (*Lepechinia calycina*), serviceberry (*Amelanchier alnifolia*), western mock-orange (*Philadelphus lewisii*).

Additional perennials to try:

Little California sunflower (*Helianthella californica*), California skullcap (*Scutellaria californica*), California milkwort (*Polygala californica*), ground iris (*Iris macrosiphon*), Purdy's iris (*I. purdyi*), California lomatium (*Lomatium californicum*), lobed violet (*Viola lobata*), goldwire (*Hypericum concinnum*).

Additional grasses to try:

Western fescue (*Festuca occidentalis*).

Additional bulbs to try:

Chaparral lily (*Lilium rubescens*), ookow (*Dichelostemma congestum*), checker lily (*Fritillaria affinis*).

Additional ferns to try:

California polypody (*Polypodium californicum*), goldback fern (*Pentagramma triangularis*), coffee fern (*Pellaea andromedifolia*).

Additional annuals to try:

Miner's lettuce (*Claytonia perfoliata*), tincture plant (*Collinsia tinctoria*), woodland nemophila (*Nemophila heterophylla*), graceful clarkia (*Clarkia gracilis* var. *sonomensis*).

Places to Visit

Ida Clayton Road, Mt. St. Helena, Sonoma and Lake counties. From Hwy 128 in Calistoga, go around four miles north to Alexander Valley and turn east onto Ida Clayton Road. This road winds around the northwestern flanks of the mountain, changes its name to Western Mines Road, and eventually intersects Hwy 29. It passes through stands of mixed-evergreen forest, oak woodland, chaparral, grassland, knobcone pine forest, and serpentine-based cypress forest. The mixed-evergreen forest includes coast live oak (*Quercus agrifolia*), California black oak (*Q. kelloggii*), canyon live oak (*Q. chrysolepis*), Garry oak (*Q. garryana*), Douglas-fir (*Pseudotsuga menziesii*), ponderosa pine (*Pinus ponderosa*), sugar pine (*P. lambertiana*), madrone (*Arbutus menziesii*), tanbark oak (*Lithocarpus densiflorus*), and California bay (*Umbellularia californica*). Understory plants include flowering dogwood (*Cornus nuttallii*), deerbrush ceanothus (*Ceanothus integerrimus*), tobacco brush (*C. velutinus*), gooseberries (*Ribes* spp.), toyon (*Heteromeles arbutifolia*), manzanitas (*Arctostaphylos* spp.), Purdy's iris (*Iris purdyi*), ground iris (*I. macrosiphon*), California milkwort (*Polygala californica*), California lomatium (*Lomatium californicum*), shooting star (*Dodecatheon hendersonii*), woodland stars (*Lithophragma* spp.), hound's tongue (*Cynoglossum grande*), Diogenes' lantern (*Calochortus amabilis*), Ithuriel's spear (*Triteleia laxa*), and lobed violet (*Viola lobata*).

Feather Falls Trail, northern Sierra, Butte County. From Oroville, take Hwy 162 east to Forbestown Road, turn right and go to Lumpkin Road, turn left and go around 12 miles to a side road on the left marked for Feather Falls. The trail, a nine-and-a-half-mile round trip, goes to the sixth highest waterfall in the United States. It passes through mixed-evergreen and conifer-

ous forest, chaparral, and riparian woodland. Trees include ponderosa pine, Douglas-fir, California bay, madrone, California black oak, tanbark oak, canyon live oak, and incense-cedar (*Calocedrus decurrens*). Understory plants include snowdrop bush (*Styrax officinalis* var. *rediviva*), western mock-orange (*Philadelphus lewisii*), western spice bush (*Calycanthus occidentalis*), whiteleaf manzanita (*Arctostaphylos viscida*), toyon, Indian pink (*Silene californica*), rainbow iris (*Iris hartwegii*), yellow pussy ears (*Calochortus monophyllus*), lobed violet (*Viola lobata*), Shelton's violet (*V. sheltonii*), Sierra fawn lily (*Erythronium multiscapoideum*), Sierra ginger (*Asarum hartwegii*), graceful larkspur (*Delphinium gracilentum*), littleleaf montia (*Montia parvifolium*), red ribbons clarkia (*Clarkia concinna*), harlequin lupine (*Lupinus stiversii*), azalea-flowered monkeyflower (*Mimulus bifidus*), and heart-leaf milkweed (*Asclepias cordifolia*).

Cone Peak, Santa Lucia Mountains, Monterey County. From the Monterey Peninsula, drive south along the Big Sur coast on Hwy 1 past the tiny town of Lucia to Ferguson-Nacimiento Road and turn left. The road takes you up and over the saddle of the Santa Lucia Mountains. At the saddle, turn left onto Cone Peak Road, which continues a few miles to the trailhead to Cone Peak. The trail passes through mixed-evergreen coniferous forest and chaparral. Trees in this area include Coulter pine (*Pinus coulteri*), sugar pine, Douglas-fir, the rare Santa Lucia fir (*Abies bracteata*), California bay, canyon live oak, coast live oak, black oak, tanbark oak, and madrone. Understory plants include Hoover's manzanita (*Arctostaphylos hooveri*), Eastwood manzanita (*A. glandulosa*), wartleaf ceanothus (*Ceanothus papillosus*), deerbrush ceanothus, canyon gooseberry (*Ribes menziesii*), golden fleece (*Ericameria arborescens*), toyon, wood rose (*Rosa gymnocarpa*), deer lupine (*Lupinus cervinus*), a rose-colored form of white globe-tulip (*Calochortus albus* var. *rubellus*), common wood fern (*Dryopteris arguta*), giant chain fern (*Woodwardia fimbriata*), modesty (*Whipplea modesta*), and hummingbird sage (*Salvia spathacea*).

Cuyamaca State Park, Laguna Mountains, San Diego County. From Julian, take Hwy S-3 south to Cuyamaca State Park. Several trails delve into the forest here. Among the conifers to be seen are ponderosa and Coulter pines, incense-cedar, bigcone Douglas-fir (*Pseudotsuga macrocarpa*), and the rare Cuyamaca cypress (*Cupressus arizonica* ssp. *stephensonii*). Broadleaf trees include coast live oak, canyon live oak, California black oak,

madrone, and bay. Understory plants include deer grass (*Muhlenbergia rigens*), serviceberry (*Amelanchier alnifolia*), squaw bush (*Rhus trilobata*), sugar bush (*Rhus ovata*), white sage (*Salvia apiana*), mugwort (*Artemisia ludoviciana*), big sagebrush (*A. tridentata*), bigberry manzanita (*Arctostaphylos glauca*), toyon (*Heteromeles arbutifolia*), Parish's blue curls (*Trichostema parishii*), and blue bush lupine (*Lupinus albifrons*).

A Douglas-fir dominates a coastal mixed-evergreen forest in Overlook Canyon in the Santa Cruz Mountains.

CHAPTER 8

OAK WOODLAND: CALIFORNIA'S SIGNATURE FOOTHILL LANDSCAPE

The broadly rounded canopies of oaks define the essence of California's foothills—rolling hills of winter-green grasses that, in the heat of summer, become richly golden. You're lucky if you live where such a landscape occurs naturally, but you can also create a small version of it in the garden.

Oak woodlands vary from place to place: near the coast, they're dominated by the deep green umbrellas of coast live oak (*Quercus agrifolia*); inland, they are likely to consist of the more open framework of blue oak (*Q. douglasii*) or the more compact shapes of interior live oak (*Q. wislizenii*). Valley bottoms and the upper shoulders of streams are shaded by the stately crowns of immense valley oaks (*Q. lobata*) that send their branchlets drooping gracefully to the ground. On uplands and to the north, stands of deciduous Garry oaks (*Q. garryana*) and massive California black oaks (*Q. kelloggii*) predominate, the latter distinguished by its flourish of bright pink new leaves in early

LEFT: Rolling hills with blue oak (*Quercus douglasii*) woodland, southern Sierra, Tulare County. • ABOVE: Valley oak (*Quercus lobata*) and masses of goldfields (*Lasthenia* sp.), Highway 58, San Luis Obispo County.

spring. In the uplands of southern California's coastal mountains, coast live oaks join company with the blue-leaved Engelmann oak (*Q. engelmannii*), which is threatened by habitat loss. At middle elevations in the southern part of the state, California black oak often occurs with pines (*Pinus* spp.), bigcone Douglas-fir (*Pseudotsuga macrocarpa*), and California bay (*Umbellularia californica*).

Sometimes these oaks are mixed with other tough, drought-tolerant trees: multitrunked, wispy-needled, gray pines (*Pinus sabiniana*) or the low, rounded canopies of the seasonally dramatic California buckeye (*Aesculus californica*). Closer to the coast or on cool, north-facing slopes, oaks mix with other evergreen broadleaf trees such as California bay, madrone (*Arbutus menziesii*), and tanbark oak (*Lithocarpus densiflorus*) to form a mixed-evergreen forest.

Oak woodlands are light-filled, open stands of trees carpeted by sweeps of native bunchgrasses studded with colorful spring wildflowers. (Oaks that are widely scattered form what is known as an oak savannah.) Under the shade of the trees themselves, look for a different set of wildflowers, together with drought-adapted ferns, shade-loving grasses, and seasonal shrubs. Despite the common belief that dry shade is difficult, California's foothills support an array of attractive plants adapted specifically to these conditions.

Because oak woodlands span a range of climatic conditions, from some summer

Oak woodland with carpet of Ithuriel's spear (*Triteleia laxa*), Table Mountain, Butte County.

fog to bone-dry summer heat, the plants that grow with them also vary. To create this habitat in the garden, pay careful attention to such local variables as slope, summer temperatures, and proximity of the water table to the surface. Visit oak woodlands in different places and at different times of the year: winter to see the deciduous oaks (blue, valley, Garry, and California black oaks) in their leafless state, spring to experience the glorious wildflower displays, and summer to appreciate oaks fully leafed out and at the height of their splendor.

Creating a Garden

The trick to creating a successful garden under oaks is to pay attention to the spacing of the trees and to work with the fact that oaks are either evergreen or deciduous. These factors determine the amount of light and shade available to understory plants, especially those that bloom early and need generous light at that time. It is also important to use understory plants that need no summer water because the surface roots of most oaks are prone to fungal infection with summer water, which dramatically shortens their life spans.

LEFT: Climbing rose on an antique gate mounted on the garage wall (featured garden). • RIGHT: Idaho fescue (*Festuca idahoensis*) and coast live oaks (*Quercus agrifolia*) in featured garden.

The interplay between dappled light and understory plants gives a special aspect to oak woodland gardens. In many cases, the best choices are drifts of herbaceous perennials, ground covers, and such adaptable grasses as California fescue (*Festuca californica*). In spring, early-blooming perennials, bulbs, and annuals further enliven the scene with a burst of color that soon gives way to the muted colors of summer dormancy.

Design Notes

This garden was designed for a planned development in Livermore featuring eclectic houses with Arts and Crafts and Mission-style architectural details, variously set back from the streets, and expansive meadow swales along the roadsides, allowing for water percolation. The houses recall a small-town American community and contribute to a defined sense of place reminiscent of an earlier time. Livermore is located some 40 miles inland from the southern edge of the San Francisco Bay. It has a rich agricultural past and is a winemaking region.

The house is a substantial Craftsman-style home with elements of American farm style, as seen in the high peaked rooflines. The views across the landscape are superb. The garden plan links the homesite to the farmlands of Livermore and the surrounding oak-studded hillsides, by means of oak woodland, grassland meadow, and chaparral plantings. Several riparian species are also planted in or near the red fescue meadow swales, which are watered in the summer.

The pool simulates a natural pond. One large western sycamore dominates the rear garden. The flagstone is used in a variety of ways: laid flat in a random pattern; stacked; tipped on edge; upright and punctuated with large boulders. The owners collect antique planters, and these we planted with exotics to provide a cottage-style garden theme. Topiaries and trellises add to this effect. On the garage wall hang three antique gates and decorative panels; from a wall planter below them rise climbing roses

A container garden with dried manzanita, juniper and cotoneaster (featured garden).

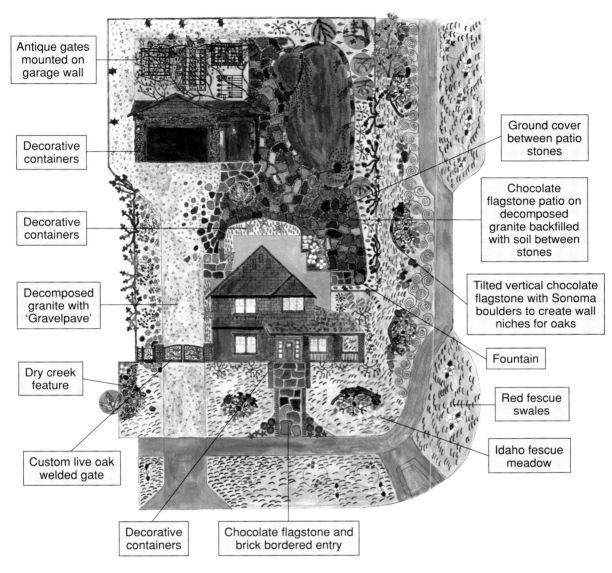

Antique gates mounted on garage wall

Decorative containers

Decorative containers

Decomposed granite with 'Gravelpave'

Dry creek feature

Custom live oak welded gate

Ground cover between patio stones

Chocolate flagstone patio on decomposed granite backfilled with soil between stones

Tilted vertical chocolate flagstone with Sonoma boulders to create wall niches for oaks

Fountain

Red fescue swales

Idaho fescue meadow

Decorative containers

Chocolate flagstone and brick bordered entry

and creeping fig. A car-park trellis and gate provide a garden-like treatment in front of the garage. The driveway is California Gold gravel, with a honeycomb plastic grid overlaid to hold the gravel in place, eliminate ruts, and provide stability for frequent vehicle use. It is also a porous surface that allows rainwater to percolate through. (This grid system, called Gravelpave, is manufactured by Invisible Structures.)

A privacy screen is achieved by planting a row of fremontodendrons, which are then espaliered to the fence. Several native bunchgrasses link these natural garden areas together. Coast live oak and acorn motifs provide art detail on gates, fences, and

Rendering of featured garden.

the flagstone wall at the pool's edge. A metal artist, David Kimpkin, created an exquisite live oak gate with a pedestrian entry using the live oak motif.

The natural vegetation of Livermore, as of many residential areas in California, is oak woodland. Unlike lawns, oaks require no summer water, and they attract more species of wildlife than any other native tree. Squirrels and blue jays provide the important benefit of burying acorns, many of which become seedling oaks. If you see one in your neighborhood, protect it. I like to plant oaks as large as the client can afford. Investing in oaks in California is investing in the natural beauty of our state.

Because oaks are fast growers (in ten years, an oak can grow ten feet), make sure young oaks get plenty of water. Deep watering once a week will reward you with abundant new growth and a healthy, robust tree that will live for several generations.

Designs for oak understory vary from massing evergreen shrubs to mixing grasses and colorful perennials, or the palette can be limited to simple evergreen ground covers. You might also try dry creek and riparian accents suggestive of water. Leaving the oak leaf litter in place contributes to the health of the oaks and the soils.

Native understory plantings for oaks can be watered during the summer months of the establishment period (one to three years). Drip irrigation during summer can also be used after establish-

Coast live oak (*Quercus agrifolia*). Illustration by A. Yankellow.

ment at infrequent intervals (once or twice a month). Extensive overhead watering under established oaks will eventually kill the oak, however, so plan the understory accordingly.

Scope of Work

- ◆ Coordinate with pool contractor to provide aesthetic transitions with flagstone.
- ◆ Demolish existing concrete at entry, driveway, and patio.
- ◆ Create a series of niches with fitted chocolate flagstone and boulders in the front garden. The grade should be altered to distribute runoff to planted areas.
- ◆ Resurface hardscape entries and porches with flagstone. Add bronze leaf details in hardscape surface.
- ◆ Install California Gold gravel and Gravelpave driveway.
- ◆ Build car-park trellis and custom gates and install.
- ◆ Install irrigation system and tie downspouts into landscaped areas with flexible drainage pipe.
- ◆ Plant nonnative ground covers between the flagstones in the pool patio garden.
- ◆ Plant containers with exotics.
- ◆ Install additional topiary.
- ◆ Install plants, as well as decorative mulch to a depth of three to four inches.

Plant List

SYMBOL	BOTANICAL NAME	COMMON NAME
A	*Quercus agrifolia*	coast live oak
B	*Cercis occidentalis*	western redbud
C	*Pinus sabiniana*	gray pine
D	*Calocedrus decurrens*	incense-cedar
E	*Leymus condensatus* 'Canyon Prince'	giant rye grass
F	*Festuca idahoensis* 'Siskiyou Blue'	Siskiyou blue fescue
G	*Heteromeles arbutifolia*	toyon
not shown	*Aesculus californica*	California buckeye
not shown	*Triteleia laxa* 'Queen Fabiola'	Ithuriel's spear
H	*Mimulus aurantiacus*	sticky monkeyflower

Chaparral Community

i	*Arctostaphylos* 'Howard McMinn'	Vine Hill manzanita

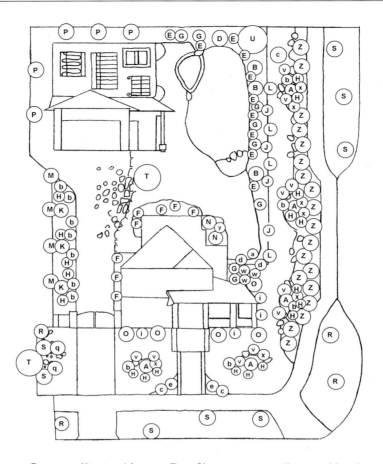

J	*Garrya elliptica* 'James Roof'	silk tassel bush
K	*Ceanothus gloriosus*	glorymat
L	*Clematis lasiantha*	virgin's bower
M	*Fremontodendron californicum*	flannelbush
N	*Carpenteria californica*	California bush anemone
O	*Arctostaphylos glauca*	bigberry manzanita

Riparian Community

P	*Vitis californica* 'Roger's Red'	California grape
q	*Juncus patens*	California rush
R	*Philadelphus lewisii*	western mock-orange
S	*Physocarpus capitatus*	ninebark
T	*Platanus racemosa*	western sycamore

Featured garden key.

Mixed-Evergreen Forest Community

| U | *Arbutus menziesii* | madrone |

Grassland Community

V	*Iris douglasiana*	Douglas iris
W	*Muhlenbergia rigens*	deer grass
X	*Festuca idahoensis*	Idaho fescue

Coastal Bluff Community

| Y | *Armeria maritima* | seathrift |
| Z | *Baccharis pilularis* 'Pigeon Point' | dwarf coyote brush |

Wetland Community (in the fountain)

a	*Anemopsis californica*	yerba mansa
a	*Ranunculus aquatilis*	water buttercup
a	*Juncus acutus*	southern rush

Coastal Sage Community

b	*Penstemon heterophyllus*	blue foothill penstemon
c	*Dudleya pulverulenta*	chalk dudleya
d	*Calystegia macrostegia* 'Anacapa Pink'	southern morning glory

Channel Islands Community

| e | *Eriogonum arborescens* | Santa Cruz Island buckwheat |

Nonnatives in containers and between stepping stones

Thymus praecox	creeping thyme
T. lanuginosus	woolly thyme
Mentha requienii	Corsican mint
Rosa 'Iceberg Standard'	iceberg rose
Rosa 'New Dawn Climber'	climbing rose
Ficus repens	creeping fig
Rosa 'Climbing Blaze'	climbing rose

Dietes spp.	fortnight lily
Juniperus nana	dwarf juniper
Thuja orientalis 'Aurea Nana'	dwarf golden Chinese thuja
Passiflora alatocaerulea	passion vine

Plants to Use

Trees

Propagation Note: All trees are easily propagated from fresh seed; oak acorns and buckeye seeds store poorly and should be immediately planted. Oaks need little maintenance once established.

Coast live oak (*Quercus agrifolia*).
Broad rounded crowns and multiple trunks 30 to 40 feet high. Bark dark gray, some-

Coast live oak (*Quercus agrifolia*) with Douglas iris (*Iris douglasiana*) and manzanita planted behind a ledgestone wall.

times fissured, often supporting lichens and mosses. Shiny, broadly elliptical, evergreen leaves with curled margins lined with prickly teeth. Male catkins are red in bud. Favors coastal hills and inland valleys of the Coast Ranges. Vulnerable to sudden oak death. Garden Design Note: Suitable primarily in coastal areas that support madrone or bay as well. If you have property with oaks present, plant an oak understory garden. If planting a coast live oak seedling, design a meadow that will transition into a woodland garden as the oak matures.

Blue oak (*Q. douglasii*).
Often as tall as broad and to 40 feet high. Small, bluish, deciduous leaves with few shallow lobes, whitish bark often longitudinally furrowed, and yellow male catkins. A well-grown specimen is strikingly beautiful. Native to hot foothills around the Central Valley in the Coast Ranges and Sierra.
Garden Design Note: This is an appropriate smaller oak for inland areas. Take the time to prune and shape it. Consider creating a habitat garden with a grove of blue oaks.

Valley oak (*Q. lobata*).
A huge, stately tree, up to 40 feet high and living as long as 300 years. Very broad, rounded crown with side branches drooping and often contorted; massive trunk with shallow, checkered, whitish bark. Large, deciduous, elliptical leaves with several deep, rounded lobes; yellow male catkins. Favors bottomlands and valleys of the inner Coast Ranges, Central Valley, and Sierra foothills.
Garden Design Note: A maturing valley oak will use 250 gallons of water a day; plant it only where the water table is high. One of the largest oaks in the world, valley oak will grow 30 to 40 feet in 20 years. Attracts the most species of wildlife.

Engelmann oak (*Q. engelmannii*).
Similar to blue oak but with unlobed leaves that are semievergreen. Now endangered in southern California due to habitat loss. Grows in mixed forests and oak woodlands in the uplands of Riverside and San Diego counties.
Garden Design Note: If you live in southern California and have the space, plant this oak. Better yet, plant a grove. Make a commitment to manage your young trees. A surrounding meadow accented with manzanita, toyon, and chaparral currant will complete the scene.

California black oak (*Q. kelloggii*).
Large trees with crowns as tall as broad, up to 50

New spring leaves of black oak (*Quercus kelloggii*) are bright pink.

feet high. Massive trunks with dark gray or black bark, deeply fissured in age. Male catkins red in bud. Large, deciduous, broadly elliptical leaves with several deep lobes, each ending in a pointed bristle. Spring leaves have velvety pink hairs; leaves turn yellow to orange in fall. Abundant at middle elevations in the north Coast Ranges, Klamath Mountains, and Sierra south to the Peninsular Ranges of southern California. Vulnerable to sudden oak death.

Garden Design Note: This is my favorite oak for home gardens, because of its year-round interest and because its size can be controlled. Great in a children's garden. Native Americans favored black-oak acorns above all others.

Gray pine (*Pinus sabiniana*).
Distinctive conifer to 60 feet high, often multitrunked. Mature bark is reddish brown and fissured. Long, sparse, drooping, grayish needles in threes are lacy and beautiful backlit. The huge, heavy seed cones with armed scales have large, tasty, nutritious seeds. Adapted to hot foothills around the Central Valley.

Garden Design Note: This pine is underutilized in suburban gardens. Create a California gray garden with 'Siskiyou Blue' fescue, blue-eyed grass, bee's bliss and California fescue as underplantings for gray pine.

California buckeye (*Aesculus californica*).
Deciduous tree with broadly rounded crown of similar shape

to live oaks, seldom topping 30 feet. Often multitrunked. Apple green, palmately compound leaves are arranged in pairs from striking buds. Fragrant, white to pale pink flowers in upright candles appear from late spring to

LEFT: Buckeye trees (*Aesculus californica*) in winter. • RIGHT: An unusual form of California buckeye (*A. californica*) with pink flowers.

early summer, attractive to many pollinators but poisonous to honey bees. Large, leathery, pear-shaped fruits in autumn split to release a single, large, shiny, chestnut-shaped seed. Beautiful smooth light gray bark and twig structure are revealed in fall and winter. Plant parts are poisonous to people. Buckeyes may be gently shaped by tip pruning and trained as an open hedge.

Garden Design Note: Outstanding winter form. Some people are bothered by its early dormancy, but it reminds us of California's natural cycle. You can manage its growth to accommodate smaller urban spaces.

Large Shrubs

Western redbud (*Cercis occidentalis*). Deciduous shrub or small tree with broad crown to 15 feet high; slow growing. Masses of magenta to rose-purple, sweet pea-like flowers appear in early to midspring just as leaves emerge. Leaves are broad, heart- to kidney-shaped, and dark green, often turning burgundy or yellow in fall. Flowers are followed by large, hanging, wine-colored, pealike pods that fade to brown. Thrives in hot foothills of the inner north Coast Ranges and Sierra. Propagate from suckers or give seeds a hot water treatment before planting. Prune out unwanted stump sprouts and suckers.

Garden Design Note: A terrific small flowering tree for the garden. Manage its growth in the early years. Can be trained as living architectural elements like arbors and arches that frame a space with vibrant color in the spring.

Blue elderberry (*Sambucus mexicana*).
Large, deciduous shrub or small tree with multiple trunks and an irregular crown reaching 20 feet high. Bark patterned similarly to oaks. Often produces "water" shoots from base, which should be pruned out. Large, pinnately compound, ill-scented leaves arranged in pairs. Flat-topped clusters of small, creamy, fragrant flowers in late spring or early summer, attractive to many pollinators. Edible, blue berries in midsummer. Adapted to canyon bottoms from the hot foothills to the middle elevations in the mountains. Propagate from suckers or stratified seed. Older plants may be renewed by coppicing the stems to the ground and allowing new sprouts to take their place.

Garden Design Note: Multifaceted small tree. Excellent habitat plant. I've used the fruit in jams, jellies, and sauces for meats, reserved the juice for sorbets, and made wine.

Flowering branches of western redbud (*Cercis occidentalis*).

Common manzanita (*Arctostaphylos manzanita*).
Large, bold, evergreen shrub to 15 feet high. Moderate
growth rate. Can be pruned to develop a single trunk.
Lustrous red-purple bark and ovate, pale to gray green,
vertically held leaves. Umbel-like clusters of fragrant
white or palest pink flowers in late winter. The edible,
reddish, applelike berries are attractive to birds and other
wildlife. Prominent throughout the inner Coast Ranges.
Propagate from hardwood cuttings in fall or from scarified or stratified seed. Common manzanita has no burl
and so should not be severely cut back.
Garden Design Note: Will live a long time but needs
excellent drainage. Create berms, or plant it on a slope
or on a mound between two large boulders. Underplant with California fescue or deer
grass.

Toyon (*Heteromeles arbutifolia*).
Large, dense, evergreen shrub or tree that may reach the size of a live oak. Brown to
gray bark; large, broadly lance-shaped, leathery, hollylike leaves; and pyramidal clus-

ters of small, white,
roselike flowers in sum-
mer. Vivid orange-red
pomes in November
and December, beauti-
ful for holiday decora-
tions. Fruits insipid but
edible and attractive to
many birds. Wide-
spread especially in
central and southern
California. Propagate
from stratified seed.
'Davis Gold' is a cultivar
with yellow fruits.
Garden Design Note:
Every oak woodland
garden deserves at least
one toyon; if you are

TOP: Flowers of common manzanita (*Arctostaphylos manzanita*). • BOTTOM: A dramatic fruiting toyon (*Heteromeles arbutifolia*) by house entry, with cast bronze and basalt bird bath sculpture by artist David Middlebrook.

lucky, one day a flock of cedar waxwings will invade and get drunk on the red berries. Very dependable and trouble free. Good screening plant. Hollywood was named for the toyon seen growing in the hills.

Small Shrubs

Chaparral currant (*Ribes malvaceum*).
Fast-growing, deciduous shrub to eight feet high with pale green, maplelike, sage-scented leaves and nodding trusses of pale pink-purple blossoms that open in winter. Insipid, pale bluish berries. Grows in open woodlands and on the edges of chaparral from central California south. Propagate from cuttings in winter or early spring. Chaparral currant can be cut back by half to keep it bushy.
Garden Design Note: This attractive winter-flowering shrub is a magnet for Anna's hummingbirds. When it's blooming, you will hear hummingbirds nearby. Use it in small groves at the edge of the garden, preferably near a window where you can watch the birds.

Fuchsia-flowered gooseberry (*R. speciosum*).
Fast-growing deciduous shrub to five feet high with arching branches and small, glossy, palmately lobed leaves. Branches are lined with spines, especially on the new growth. Lines of long, tubular, bright red, fuchsialike flowers, appearing from late winter to midspring, are attractive to hummingbirds. Spiny fruits. Grows in open woodlands from central California south. Propagate from cuttings in winter and early spring. Prune out old, dead canes.
Garden Design Note: An oak woodland garden with an Asian theme can feature this gooseberry prominently. Shape it to fit the space. Arching branches on a slope are lovely, or display it with selected rocks in a Chinese pot at the entry to your home.

Blue-witch (*Solanum umbelliferum*).
Semievergreen to summer-deciduous, multibranched shrub to three feet tall with intricate greenish to silvery twigs. Small, elliptical, pale green leaves lost when summers are hot and dry. Dense clusters of fragrant, sky-blue, saucer-shaped flowers centered with yellow stamens in early to midspring. Propagate from stratified seed or cuttings.

Flowering branches of fuchsia-flowered gooseberry (*Ribes speciosum*).

Garden Design Note: Plant blue-witch with narrow-leaved goldenbush for spring color: they bloom at the same time. Good drainage and plenty of sun will assure the best performance.

Narrow-leaved goldenbush (*Ericameria linearifolia*).
Low, rounded, evergreen shrub to three feet tall with glossy, fragrant, linear leaves. Showy, bright yellow, margueritelike flowers in midspring. Widespread in dry inner foothills and into desert mountains. Propagate from seed.
Garden Design Note: An attractive yellow flowering bush for the garden. Makes a fine accent in a meadow planting; use several to punctuate a grassy hillside. May be difficult to find. Needs excellent drainage.

Bush snowberry (*Symphoricarpos albus* var. *laevigatus*).
Colonizing, deciduous shrub to four feet high with arching, finely textured twigs and pairs of rounded to elliptical, pale green leaves. Leaf margins irregularly lobed or toothed. Tiny, pink, bell-shaped flowers appear under leaves near twig tips in late spring to summer, followed by spongy, soapy, inedible white berries in fall. Widespread in forests and woodlands. May be invasive. Propagate from divisions.
Garden Design Note: Likes partial shade. Plant under mature oaks. 'Tilden Park' is an excellent selection. There is also a creeping form of snowberry, *Symphoricarpos mollis*.

Perennials and Grasses

California fescue (*Festuca californica*).
Medium-sized bunchgrass to three feet high with fountain-like clusters of narrow green to bluish green, recurved leaves. Arching panicles of spikelets in late spring. Occasional summer water best. Propagate from divisions or seed. Remove the old thatch every winter to encourage new growth in spring.

Garden Design Note: This grass, one of the most dependable to plant under oaks, can take sun or shade. 'Blue Fountain', selected by Nevin Smith, and, for southern California gardens, 'Horse Mountain Green', are both good.

Flowering plant of California fescue (*Festuca californica*).

Mule's ears (*Wyethia* spp.).
Winter-dormant, rhizomatous plants with large basal leaves that vary from wavy and lance-shaped to flat and broadly ovate. Leaves are grayish or bright green. Large, four-inch-broad, golden flower heads resembling sunflowers bloom mid to late spring. Attractive to bees. Propagate from divisions or seed.
Garden Design Note: This large perennial with straplike leaves and big flowers looks great in a hillside meadow. Difficult to find.

Shooting star (*Dodecatheon hendersonii*).
Summer/fall-dormant from fleshy roots with bulblets. Rosettes of oval leaves in winter are followed by naked scapes bearing umbels of cyclamenlike, pink to rose-purple flowers in early spring. Each blossom has swept-back petals and a cone of stamens

Masses of California fescue (*Festuca californica*) under manzanitas (*Arctostaphylos* spp.).

that shoot forward. Plant in drifts. May be treated like bulbs. Propagate from divisions or seed; seedlings take a couple of years to reach flowering size.

Garden Design Note: Sometimes you can find these at specialty nurseries that sell bulbs.

Hound's tongue (*Cynoglossum grande*).
Summer/fall-dormant plants from chunky rootstocks. In winter, bronze-tinted leaves unfurl into green, broadly tongue-shaped blades. Flowering stalks to three feet high in early spring carry clear blue, forget-me-not-like flowers with a central white corona. Wart-covered, one-seeded nutlets. Propagate from fresh, stratified seed.

Garden Design Note: I have purchased one or two of these plants at California Native Plant Society sales, but production nurseries are not growing it. I would love to plant 20 along a path in an oak woodland garden, but alas, I can't find them. Not yet, anyway.

California buttercup (*Ranunculus californicus*).
Short-lived, summer/fall-dormant perennial from fleshy roots. Most leaves are basal

and divided into several deep palmate lobes. Shiny, varnished yellow flowers appear from late winter to midspring in open panicles on stalks to a foot high. May reseed. Propagate from seed. Restart often.

Garden Design Note: Charming perennial that can be treated like an annual. Mix with lupine, delphinium, and owl's clover.

Hummingbird sage (*Salvia spathacea*).
Perennial ground cover that spreads by rhizomes to create large mats several feet across and up to three feet high in flower. Sweetly fragrant, triangular-elliptical leaves, with stout spikes of rose-pink flowers attractive to hummingbirds in spring. May die back without summer water. Propagate from divisions or seed.

TOP: Flowers of Henderson's shooting star (*Dodecatheon hendersonii*). • BOTTOM: Flowering spike of hummingbird sage (*Salvia spathacea*).

Garden Design Note: These are good blooming plants under oaks. They will spread, but slowly, so mass for dramatic effect if you can't wait for them to colonize. Mix with California fescue and Sonoma sage (*Salvia sonomensis*) in sunny openings.

Scarlet larkspur (*Delphinium nudicaule*).
Short-lived, summer/fall-dormant perennial from thick rootstock. Mostly basal leaves are rounded and deeply palmately lobed. Racemes of scarlet, spurred flowers appear on stalks to two feet high in midspring. Propagate from seed. Needs excellent drainage and protection from snails and slugs. In a sunny place like southern California, use *D. cardinale*.
Garden Design Note: Thompson and Morgan Nursery in England has a selection 'Laurin' that is very gardenworthy.

Western or foothill wallflower (*Erysimum capitatum*).
Biennial or short-lived perennial from taproot to four feet high (often less) with narrowly lance-shaped leaves and tight racemes of fragrant, yellow to burnt orange flowers. Blooms mid to late spring. Propagate from seed. Protect from snails and slugs.
Garden Design Note: A unique shade of yellow-orange, the flowers are massed at the top of the stalk. A stunning combination with elegant clarkia and red maids.

Hummingbird sage (*Salvia spathacea*) with Cleveland sage (*Salvia clevelandii*) and ceanothus.

Ferns

Common wood fern
(*Dryopteris arguta*).
Medium-sized, ever-green, clumping fern with twice-divided feathery fronds to two feet high. Sori circular and covered by a horseshoe-shaped indusium. No summer water. Remove dead fronds at year's end. Garden Design Note:

Mix with masses of lady fern and western sword fern; punctuate with snowberry.

Bulbs

Note: Bulbs should be lifted and stored in a cool place if the garden receives a lot of summer water.

White fairy lantern or globe-tulip (*Calochortus albus*).
Foot-tall stalks from a single, basal, strap-shaped leaf. Stalks bear up to a dozen exquisite, white, baublelike globes—sometimes flushed rose, sometimes pure white—in midspring. Prefers steep, well-drained banks. Stunning in large colonies. Propagate from bulblets or seed.
Garden Design Note: Available at rare bulb purveyors like Telos Rare Bulbs. Sometimes you can find a pink form. Loves a slope and excellent drainage.

Ithuriel's spear (*Triteleia laxa*).
Six-inch- to two-foot-tall stiff naked flowering stalks bear open umbels of sky blue to deep blue-purple, funnel-shaped flowers resembling aga-panthus blossoms. Basal leaves often shrivel by flowering time. Beautiful in large drifts, especially with yellow mariposa-tulips and wind poppies. Propagate from cormlets or seed.
Garden Design Note: Gophers and deer consider these plants delicacies. This is the

TOP: Common wood fern (*Dryopteris arguta*) surrounded by common wood mint (*Stachys ajugoides*). • BOTTOM: Flowers of white fairy lantern (*Calochortus albus*).

easiest bulb to grow, and it will tolerate clay. Plant many (over 200) for the best effect. Mix with poppies, lupines, and clarkias for a dependable, showy meadow.

Annuals

Propagation Note: All annuals are propagated from seed.

Chinese houses (*Collinsia heterophylla*). Plants, which reach 12 inches in height, form bushy clumps when given extra water. In midspring, tiers of pagodalike, two-lipped blossoms emerge, each with a near-white upper lip and rich purple lower lip. Plant in large drifts. Often reseeds. Grow with red ribbons, elegant clarkia, and wind poppy.

Garden Design Note: Can tolerate some shade. Stunning selections such as *Collinsia* Bicolour Surprise Mix are available through Thompson and Morgan in England.

Wind poppy (*Stylomecon heterophylla*). Basal, pinnately slashed leaves. Stems to eight inches high carry a single, large, saucer-shaped, deep orange flower with a dark purple center. Shakerlike seed pods. Responds like Chinese houses to extra moisture. Collect seeds to replant next season.

Garden Design Note: Also tolerates some shade. A woodland annual. Difficult to establish. Reseed in subsequent years. Underutilized.

Elegant clarkia (*Clarkia unguiculata*). Grows to three to four feet tall in rich, moist, well-drained soils. The young ovate leaves are often decorated with red-purple veins. Pink-purple, fanlike flowers in slender, elegant racemes appear in late spring and early summer. Reseeds. This is the parent of the hybrid clarkias available in commercial seed packets.

TOP: Field of Ithuriel's spear (*Triteleia laxa*). • BOTTOM: Flowering spike of Chinese houses (*Collinsia heterophylla*).

Garden Design Note: Many different hybrids are available, for example 'Apple Blossom', 'Royal Bouquet Mixed', and a compact form found in Santa Cruz County that makes stocky columnar plants with bright mauve flowers.

Fiesta flower (*Pholistoma auritum*).
Scrambling, weak-stemmed plants to three feet tall with coarsely pinnately lobed leaves, stems lined with recurved prickles, and clusters of deep purple, saucer-shaped flowers with a dark center. Flowers look like baby-blue-eyes but have a richer color. Place where stems can scramble up small shrubs. Dramatic in masses in early to midspring.
Garden Design Note: This plant is very effective rambling through grasses with California buttercups and blue-eyed grass.

Fernleaf phacelia (*Phacelia distans*), Douglas iris (*Iris douglasiana*), and *Heuchera* 'Wendy' under California black oaks (*Quercus kelloggii*).

CLOCKWISE FROM TOP: Oak woodland planting in a shopping center with hollyleaf cherry (*Prunus ilicifolia*) and flat-topped buckwheat (*Eriogonum fasciculatum*). • Oak woodland-grassland garden with Idaho fescue (*Festuca idahoensis* 'Siskiyou Blue') planted among ceramic vases. Photograph by S. Ingram. • California quail mosaic container by artist Christina Yaconelli. • New spring leaves on snowberry (*Symphoricarpos albus* var. *laevigatus*).

TOP: Featured oak woodland-grassland garden with a driveway gate showing oak branch motif. Artist: David Kimkin. • BOTTOM LEFT: Dwarf coyote brush (*Baccharis pilularis*) in an oak woodland garden with urbanite patio. • BOTTOM RIGHT: Dry creek in an oak woodland garden.

Additional oaks to try:

Garry oak (*Quercus garryana*), canyon live oak (*Q. chrysolepis*).

Other trees to plant with oaks:

California bay (*Umbellularia californica*), madrone (*Arbutus menziesii*).

Additional large shrubs to try:

Hollyleaf redberry (*Rhamnus ilicifolia*), deerbrush (*Ceanothus integerrimus*), wart-leaf ceanothus (*C. papillosus*), hollyleaf cherry (*Prunus ilicifolia*), hopbush (*Ptelea crenulata*).

Additional small shrubs to try:

Canyon gooseberry (*Ribes menziesii*), bush anemone (*Carpenteria californica*), western mock-orange (*Philadelphus lewisii*), creambush (*Holodiscus discolor*), Lemmon's ceanothus (*Ceanothus lemmonii*), littleleaf ceanothus (*C. foliosus*).

Additional perennials to try:

Balsamroot (*Balsamorrhiza deltoidea*), ground iris (*Iris macrosiphon*), Indian pink (*Silene californica*), showy phlox (*Phlox speciosa*), Oregon sunshine (*Eriophyllum lanatum*).

Additional bulbs to try:

Twining brodiaea (*Dichelostemma volubile*), firecracker brodiaea (*D. ida-maia*), rose globe-tulip (*Calochortus amoenus*), golden brodiaea (*Triteleia ixioides*), crinkled onion (*Allium crispum*).

Additional annuals to try:

Woodland nemophila (*Nemophila heterophylla*), tincture plant (*Collinsia tinctoria*), bride's bouquet (*Collomia grandiflora*), five-spot (*Nemophila maculata*), globe gilia (*Gilia capitata*).

Valley oak acorn finial by artist Ben Hunt.

Places to Visit

Mendocino Pass, Mendocino and Glenn counties. From Covelo, take Mendocino Pass Road to the top of the 5,000-foot-high pass, proceeding through oak woodland, mixed-evergreen and coniferous forests, and grassland along the way. The oaks vary from place to place. At the start, in Round Valley, there are magnificent valley oaks (*Quercus lobata*). The foothills beyond are home to interior live and blue oaks (*Q. wislizenii* and *Q. douglasii*). Above is a forest of madrone (*Arbutus menziesii*), Douglas-fir (*Pseudotsuga menziesii*), ponderosa pine (*Pinus ponderosa*), incense-cedar (*Calocedrus decurrens*), canyon live oak (*Q. chrysolepis*), and California black oak (*Q. kelloggii*). At the pass is extensive Garry oak (*Q. garryana*) woodland with an admixture of black oak. Plants in the understory include firecracker brodiaea (*Dichelostemma ida-maia*), various clarkias (*Clarkia* spp.), bride's bouquet (*Collomia grandiflora*), goddess mariposa-tulip (*Calochortus vestae*), giant trillium (*Trillium albidum*), checker-lily (*Fritillaria affinis*), scarlet fritillary (*F. recurva*), showy phlox (*Phlox speciosa*), California fawn-lily (*Erythronium californicum*), ookow (*Dichelostemma congestum*), and western bleeding heart (*Dicentra formosa*).

Loafer Creek State Park, Lake Oroville, Butte County. The park is located off Hwy 62 about seven miles east of Oroville on Lake Oroville. It is a fine example of blue oak woodland mixed with gray pines (*Pinus sabiniana*) and scattered interior live oaks. Understory plants include white globe-tulip (*Calochortus albus*), whiteleaf manzanita (*Arctostaphylos viscida*), buckbrush (*Ceanothus cuneatus*), lupines (*Lupinus* spp.), checkerbloom (*Sidalcea malviflora*), Ithuriel's spear (*Triteleia laxa*), Venus mariposa-tulip (*Calochortus venustus*), white brodiaea (*Triteleia hyacinthina*), and clarkias.

Morgan Territory Regional Park, Contra Costa County. From Hwy 580 take the North Livermore Road exit and drive north a few miles. Turn north (right) onto Morgan Territory Road and continue five miles to the main parking lot on the east side of the road. Several trails leave from here, passing through riparian woodland, oak woodland, chaparral, and grassland. The oak woodlands are particularly varied, with coast live oak (*Quercus agrifolia*), valley oak, California black oak, blue oak, interior live oak, and canyon live oak on different exposures. Neighboring trees include gray pine, bigleaf maple (*Acer macrophyllum*), California bay (*Umbellularia californica*), and western sycamore (*Platanus racemosa*). Understory plants include checker-lily, Mt. Diablo globe-tulip (*Calochortus pulchellus*), clarkias, sweet-cicely (*Osmorhiza chilensis* and *O. brachypoda*), California polypody fern (*Polypodium californicum*), common wood fern (*Dryopteris arguta*), California maidenhair (*Adiantum jordanii*), shoot-

Oak woodlands in spring.

ing star (*Dodecatheon hendersonii*), hound's tongue (*Cynoglossum grande*), Ithuriel's spear, mule's ears (*Wyethia* spp.), common manzanita (*Arctostaphylos manzanita*), canyon gooseberry (*Ribes menziesii*), creambush (*Holodiscus discolor*), and toyon (*Heteromeles arbutifolia*).

Hunter-Liggett area, Monterey County. Take the Jolon Road exit from Hwy 101 just north of King City. After going over a pass, the road enters the Hunter-Liggett Military Reservation, passing by the site of the former town of Jolon. The area is home to immense valley oaks, while the surrounding hills support a woodland of blue oaks and gray pines. In a good spring the whole area is ablaze with annual wildflowers, including California buttercup (*Ranunculus californicus*), royal larkspur (*Delphinium variegatum*), redmaids (*Calandrinia ciliata*), popcorn flower (*Plagiobothrys nothofulvus*), goldfields (*Lasthenia* sp.), tidy-tips (*Layia platyglossa*), owl's clovers (*Castilleja exserta* and *C. densiflora*), golden stars (*Bloo-*

meria crocea), Jolon brodiaea (*Brodiaea jolonensis*), blue dicks (*Dichelostemma capitatum*), purple amole (*Chlorogalum purpureum*), California poppy (*Eschscholzia californica*), and blue-eyed grass (*Sisyrinchium bellum*).

The highlands near Julian, San Diego County. The Julian area in the mountains west of Anza-Borrego State Park is where highways 79 and 78 meet. The area includes mixed conifer-broadleaf forest and oak woodland with ponderosa pine, Coulter pine (*Pinus coulteri*), bigcone Douglas-fir (*Pseudotsuga macrocarpa*), madrone, California bay, Engelmann oak (*Quercus engelmannii*), coast live oak, California black oak, canyon live oak, and more. Understory plants include red bush monkeyflower (*Mimulus puniceus*), southern phlox (*Phlox austromontana*), toyon, common wood fern, southern mugwort (*Artemisia ludoviciana*), golden yarrow (*Eriophyllum confertiflorum*), and common wild sweet pea (*Lathyrus vestitus*).

Oak woodlands, Morgan Territory Regional Park, Contra Costa Co.

CHAPTER 9

GRASSLANDS: PARADISE FOR WILDFLOWERS

A large part of California's hills, valleys, and mountains are covered with grasses, creating a showcase for unrivaled seasonal displays of colorful wildflowers. In this chapter we will explore summer-dry grasslands, which, found throughout California's foothills—extending from coastal bluffs across the Central Valley to the western Sierra—are characterized by spring wildflowers and summer dormancy. Summer-wet grasslands, featuring summer wildflowers and winter dormancy, occur in California as well, but at higher elevations: they are the classic montane and alpine meadows, which are detailed in Chapter 6.

Summer-dry grasslands include coastal prairies with summer fog and valley grasslands with summer sun. Although many wildflowers live under both conditions, species from coastal grasslands better serve coastal garden sites, while species from interior valleys thrive best beyond the fog belt. The bunchgrasses that form the backbone of grasslands, whether summer sunny or foggy, grow best on deep soils where their long, fibrous roots can reach deep for nutrients and recycle them back to the topsoil. Grassland soils are typically adobe clays or silts.

Most of our native grasslands, which were defined by clumped, perennial bunchgrasses, have been irrevocably changed by the introduction of alien species. The grasslands you see in agricultural and urban areas today are dominated by annual, nonnative grasses that grow and die with the rains. Bunchgrasses, in contrast, furnish year-round texture and color—green in winter and spring, golden in summer and fall—with ample space

LEFT: Field of owl's clover (*Castilleja* sp.) and dove lupine (*Lupinus bicolor*). • ABOVE: Overview of valley grassland in flower, Bear Valley.

for annual and perennial wildflowers, and bulbs. They are a wonderful focus for a garden design.

Creating a Garden

Besides the grass framework, the element that brings great drama to foothill meadows is the long succession of wildflowers—perennials, annuals, and bulbs—that bloom as early as March and continue well into June. Careful garden planning will guaran-

TOP: Antelope Valley Poppy Reserve with goldfields (*Lasthenia* sp.) and sky lupine (*Lupinus nanus*). • BOTTOM: Flowers of tidy-tips (*Layia platyglossa*).

tee long bloom, with occasional summer water extending the bloom period and pro-longing the time when grasses remain green. (Some bunchgrasses go dormant no mat-ter what the conditions are.)

Even though grasses predominate in true foothill grasslands, sedges (*Carex* spp.) may be substituted to create a meadow in the home garden. Many sedges are green through summer, and several are green year round. Be sure to pay particular attention to the ultimate size of the plant—sedges range from a few inches to six feet high—and to whether they're clump forming with short rhizomes or produce long, invasive rhi-zomes. Two excellent choices for gardens are Berkeley sedge (*C. tumulicola*) and *C. pansa*.

The selection of material for both the grass framework and the wildflower displays should be chosen according to presence of summer fog versus hot summer sun. Note that the majority of perennials favor coastal conditions, while many annuals prefer warm, inland sites. If you're incorporating bulbs into your grassland meadow, be sure to choose bulbs that are compatible with summer water if you want a green meadow, or, if you elect to let your meadow go brown, find bulbs with a definite dormancy.

Most meadows benefit visually from a background of woody shrubs, which help

Featured garden with native bunchgrasses and poppies.

define the meadow's space and borders. Appropriate shrubs may be selected from the chaparral palette, for inland sites, or coastal sage scrub for coastal sites. Examples include various wild lilacs, manzanitas (*Arctostaphylos* spp.), western redbud (*Cercis occidentalis*), and fremontia (*Fremontodendron* hybrids) for summer-hot meadows. For coastal meadows, try coffeeberry (*Rhamnus californica*), laurel sumac (*Malosma laurina*), lemonade berry (*Rhus integrifolia*), coastal ceanothus, and various sages (*Salvia* spp.).

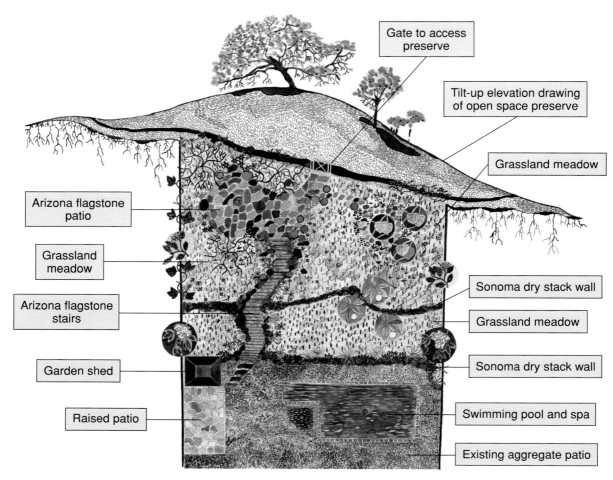

Design Notes

The house for which this garden was created lies on the outskirts of the Almaden Valley in Santa Clara County. The area designated for the native garden was originally an easement controlled by the county and maintained as a firebreak between the sub-

Rendering of featured garden.

urban housing development and an open-space preserve. The county recently relinquished rights to this easement and informed the owners that they could use the property as their own, as well as access a bordering trail.

The design challenge was to integrate the Mediterranean stucco and wood frame architectural style of this housing development with the oak woodland hillside of the open-space preserve, unifying it as one natural place.

One of our goals was to use hardscape materials native to the site or very similar in appearance. At the upper edge of the property, for example, we installed a patio using Arizona flagstone, which is comparable in color to the sandstone that was quarried from nearby hills to build Stanford University. The fountain and birdbath are both natural stone quarried locally. The upper and lower patios are linked by a curving flagstone stairway, while dry stacked stone walls weave in and out along the natural contour of the hillside.

Plant selections for this garden were derived from the oak woodland and grassland communities found locally, using the principles of succession. On the hillside facing the house (southeast), chaparral plants now drape over the existing slump-stone wall. This area is very hot and unforgiving in the summer; it also has an impervious patio surface that reflects heat. Eventually the chaparral species will be succeeded by oak woodland species when newly planted trees—oaks, buckeyes, and redbuds—create the necessary shade for understory plants to thrive.

To begin the process of reestablishing a healthy ecosystem, the majority of the garden was planted with native California bunchgrasses—California fescue, Idaho fescue, foothill needlegrass, and California melic—that might have grown here before grazing and the introduction of more competitive European annual grasses. With the grasses, we planted native flowering annuals, perennials, and bulbs. The color bonanza of the wildflower meadow features California poppies, wind poppies, elegant clarkia, red ribbons, farewell-to-spring, lupine, tidy-tips, creamcups, and globe gilia. These annual seeds were sown in drifts, as they occur in nature. The perennials include Douglas iris, royal larkspur, yarrow, checkerbloom, hounds tongue, and soapplant, while the bulbs are pink meadow onion, Ithuriel's spear 'Queen Fabiola', and yellow mariposa-tulip.

In addition to the colorful wildflower meadow, the showy parts

TOP: Yarrow (*Achillea millefolium*). • BOTTOM: Idaho fescue (*Festuca idahoensis* 'Siskiyou Blue'). Illustrations by A. Yankellow.

of the garden include a penstemon walk featuring six out-standing native California penstemon species. These are tucked among the rocks and boulders that line the stair-way, so are easily visible from the house. They were comple-mented by Santa Lucia monkeyflower (a large azalealike mimulus indigenous to the Santa Lucia Range), yarrow, and live-forever succulents commonly found hidden among the rocky hillsides of Santa Clara County.

Another dramatic surprise greets guests on the upper patio in late summer, in the form of two species of buckwheat that flower in summer and retain their orange to rust-colored flowers until December. These long-lived pe-rennials show off especially well with blue and purple asters and red California fuch-sias, which are likewise radiant in late summer.

Establishing a native bunchgrass and wildflower meadow requires at least three years of diligent weed pulling and reseeding in sections that take a while to germinate well. After that, you will be rewarded with an authentic California meadow that will outcompete invasive weed species and, with the addition of appropriate perennials, produce abundant color from spring into late fall.

TOP: Hillside meadow with native bunchgrasses, vineyard and chaparral planting. • BOTTOM: Idaho fescue (*Festuca idahoensis*) meadow punctuated with California rushes (*Juncus patens*).

Scope of Work

- ◆ Remove all debris and specified plants.
- ◆ Build a dry stacked stone wall (Sonoma fieldstone or California granite: match stone with existing stone on site, to the extent possible), not to exceed 18 inches high, following the natural contours of the hillside.
- ◆ Build a stepping-stone path of random-cut flagstone (again matching existing stone on site).
- ◆ Create a random-cut Arizona flagstone patio on the upper level of the easement.
- ◆ Install large boulders at the edges of the upper patio, as well as at selected sites throughout the garden.
- ◆ Create tree wells or small retention walls as needed for large trees.
- ◆ Install stone fountain.
- ◆ Install electricity to the upper patio and six low-voltage lights.
- ◆ Create a series of native Californian community-themed gardens, including oak woodland, grassland/meadow with wildflowers, chaparral border, and penstemon walk.
- ◆ Overhead spray irrigation will be designed and installed for a three-year operation period, giving maximum attention to the optimum establishment of the trees, grassland, and the wildflower meadow.
- ◆ Long-term drip irrigation on a separate valve will be required for containers and along the penstemon walk and the borders of the upper patio as replacements will be made from time to time if perennial color is to be a seasonal focal point.

Plant List

Rear Garden: Grassland Meadow, Oak Woodland, and Chaparral

SYMBOL	BOTANICAL NAME	COMMON NAME
e	*Achillea millefolium*	(white) yarrow
B	*Aesculus californica*	California buckeye
C	*Arctostaphylos uva-ursi*	kinnikinnick
J	*Aster chilensis*	coast aster
E	*Calystegia purpurata* 'Bolinas'	pink morning glory
b	*Ceanothus thyrsiflorus* 'Skylark'	coast blueblossom
G	*Cercis occidentalis*	western redbud
Y	*Clematis lasiantha*	virgin's bower
F	*Dudleya cymosa, D. caespitosa*	mixed live-forevers
D	*Epilobium canum*	California fuchsia

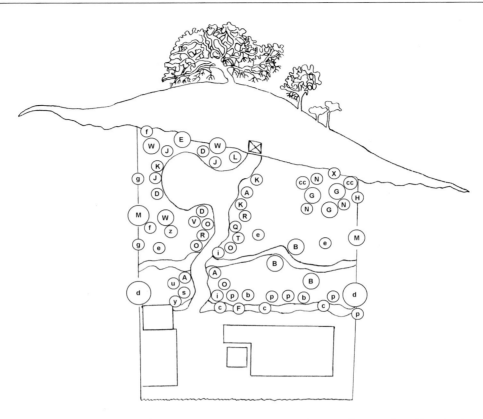

K	*Eriogonum arborescens*	Santa Cruz Island buckwheat
u	*E. fasciculatum*	California buckwheat
M	*Heteromeles arbutifolia*	toyon
cc	*Iris douglasiana*	Douglas iris
T	*Mimulus bifidus* ssp. *fasciculatus*	Santa Lucia monkeyflower
V	*Muhlenbergia rigens*	deer grass
Q	*Penstemon azureus*	azure penstemon
R	*P. centranthifolius*	scarlet bugler
O	*P. heterophyllus* var. *australis*	blue foothill penstemon
A	*P. palmeri*	Palmer's penstemon
i	*P. rydbergii*	meadow penstemon
s	*P. spectabilis*	showy penstemon
W	*Quercus agrifolia*	coast live oak
X	*Rhamnus californica*	coffeeberry
not shown	*Ribes sanguineum* var. *glutinosum*	pink-flowering currant
j	*R. speciosum*	fuchsia-flowered gooseberry

Featured garden site plan.

z	*Rosa gymnocarpa* or *R. woodsii* var. *ultramontana*	wood rose
p	*Salvia leucophylla* 'Pt. Sal'	purple sage
not shown	*S. sonomensis* 'Dara's Choice'	Sonoma sage
N	*S. spathacea*	hummingbird sage
d	*Sambucus mexicana*	blue elderberry
L	*Sisyrinchium bellum*	blue-eyed grass
f	*Symphoricarpos albus* var. *laevigatus*	snowberry
H	*Trichostema lanatum*	woolly blue-curls
g	*Vitis californica* 'Roger's Red'	California wild grape

Meadow Plantings from Seeds and Gallon Containers

Grasses

Festuca californica	California fescue
Nassella pulchra	purple needlegrass
Melica californica	California melic

Bulbs

Allium unifolium	paper onion
Triteleia laxa 'Queen Fabiola'	Ithuriel's spear
Dichelostemma ida-maia	firecracker brodiaea

Annuals from Seed

Eschscholzia californica	California poppy
Stylomecon heterophylla	wind poppy
Clarkia unguiculata	elegant clarkia
C. concinna	red ribbons
C. amoena	farewell-to-spring
Nemophila maculata	five-spot

Perennials from Seed

Cynoglossum grande	hound's tongue
Calandrinia ciliata	redmaids
Lupinus nanus	sky lupine
Layia platyglossa	tidy-tips
Platystemon californicus	creamcups

Perennials

Sidalcea malviflora	checkerbloom
Chlorogalum pomeridianum	soapplant

Plants to Use

Bunchgrasses for the Fog Belt

Propagation Note: Bunchgrasses are easily propagated by division or seed.

Nutka reed grass (*Calamagrostis nutkaensis*).
Large bunchgrass with moderately wide, bright green blades and flowering culms to four feet high. Tolerates some shade. Large, dramatic plumes of spikelets in late spring. Requires summer water. Cut out old canes each year.
Garden Design Note: This large grass can be used to punctuate a *Carex pansa* meadow. Also effective in a woodland meadow with some shade inland.

Leafy reed grass (*C. foliosa*).
Graceful clumps of narrow, pale green leaves and dense, feathery panicles of florets to around 12 inches high in late spring. Native to the north coast. Remove old thatch each winter and cut back old flower plumes.
Garden Design Note: This grass looks good when planted as a monoculture 24 inches apart. Its tawny brown flower spikes are striking, especially from above. One of our most attractive grasses in flower, it likes a coastal influence or shade inland.

Tufted hairgrass (*Deschampsia caespitosa*).
Robust bunchgrass with stiff green leaves and flowering culms to three feet high. Spikelets grow in loose, open to rather compressed panicles, the flowers with hairlike awns. Maintain as for leafy reedgrass.
Garden Design Note: Subspecies *caespitosa* is rather tolerant of dry shade situations, as under oaks, but it will go dormant in summer. Mix it with dwarf coffeeberry and hummingbird sage.

Masses of leafy reedgrass clumps (*Calamagrostis foliosa*) create drama in gardens.

Red fescue (*Festuca rubra*).
A modest-sized, rhizomatous bunchgrass that forms a turf with narrow leaves. Flowering culms to a foot and a half high, sometimes tinged purple to reddish, are in panicles. Red fescue lives in many different habitats, and consequently can be used in a wide variety of sites. Choose material according to whether it is clumping or turf-forming, and according to its origin.

Garden Design Note: Selection 'Jughandle' forms small, stiff clumps of steel-gray foliage. 'Molate' is a drought-tolerant, dark green, creeping grass with fine-textured leaves and a soft mounding habit. Once established, red fescue may crowd out annuals. Effective as a meadow grass and a lawn substitute.

Bunchgrasses for Hot Summers

Note: All large bunchgrasses should have the dead thatch removed at year's end. The most robust can also be cut back in winter to promote new growth.

Deer grass (*Muhlenbergia rigens*).
Large, robust bunchgrass with curved, narrow, light green blades and stiff, narrow,

Leafy reedgrass (*Calamagrostis foliosa*) in flower with dune tansy (*Tanacetum camphoratum*) and low growing manzanitas (*Arctostaphylos* spp.).

four-foot-tall flowering spikes. Good for large-scale plantings. Garden Design Note: This grass is too large for expansive meadow plantings except as an accent. Lends its form to container gardens, architectural plantings, and dramatic accents.

Needlegrasses (*Nassella* spp.). Modest, medium-sized bunch-grasses, some species of which once occupied a broad range. Leaves dull to gray-green with flowering culms to two or three feet high depending on species. Spikelets in open, airy panicles with a narrow outline and distinctive bent or twisted awns extending far beyond the grains. Good basic grasses for dry meadows; choose species according to locality. Attrac-

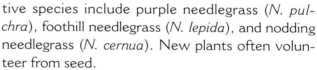

tive species include purple needlegrass (*N. pulchra*), foothill needlegrass (*N. lepida*), and nodding needlegrass (*N. cernua*). New plants often volunteer from seed.

Idaho fescue (*Festuca idahoensis*). Modest-sized bunchgrass with very narrow leaves, often of a striking blue- to gray-green color. Flowering culms from 12 to 18 inches high. Good for texture/color contrasts. (Although the cultivar

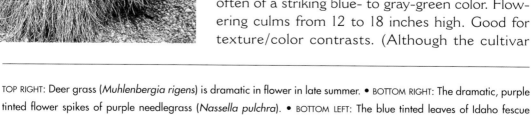

TOP RIGHT: Deer grass (*Muhlenbergia rigens*) is dramatic in flower in late summer. • BOTTOM RIGHT: The dramatic, purple tinted flower spikes of purple needlegrass (*Nassella pulchra*). • BOTTOM LEFT: The blue tinted leaves of Idaho fescue (*Festuca idahoensis*) at Regional Parks Botanic Garden.

'Siskiyou Blue' is now thought to derive from a nonnative species, several forms of Idaho fescue do have distinctly blue foliage.)

Garden Design Note: For warmer inland meadows, Idaho fescue is an excellent choice. It can take more heat than red fescue, and annuals will reseed in the openings between its foot-wide clumps. Mix red fescue and Idaho fescue for a verdant meadow.

Squirreltail grass (*Elymus elymoides,* formerly known as *Sitanion hystrix*).
Low, short-lived bunchgrass, best treated as an annual. Leaves gray-green and soft; flowering culms to two feet high with dense, brushy spikes of spikelets. The spikes are narrow and reddish in flower, later opening into a bushy "squirreltail" in fruit, when the many parts separate and are carried away by wind.

Garden Design Note: A novelty grass attractive for its fruit. Use in containers with colorful perennials.

Massed Idaho fescue (*Festuca idahoensis* 'Siskiyou Blue') at the entry to a contemporary home.

Perennial Wildflowers

California poppy (*Eschscholzia californica*).

California's state flower is a short-lived perennial or an annual featuring tufts of bluish green, ferny foliage and a very long succession of vivid orange to golden, cup-shaped flowers. Reseeds freely. Cut back periodically for better shape and longer bloom. Avoid transplanting, since the taproots are easily

damaged. White-, pink-, and red-flowered cultivars are available.

Garden Design Note: Hybrids include 'Apricot Flambeau', 'Buttermilk', 'Champagne and Roses', 'Strawberry Fields', 'Summer Sorbet', and 'Thai Silk Lemon Bush'. Use in a successional garden with grasses and other annuals. Instant gratification—almost.

Blue-eyed grass (*Sisyrinchium bellum*).

Small, tufted, rhizomatous plants with miniature fans of iris-like foliage. Clusters of blue-purple, saucer-shaped flowers

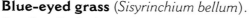

TOP RIGHT: California poppies (*Eschscholzia californica*) in a Mediterranean/cottage-style garden. Photograph by S. Ingram. • LEFT: Grassland meadow with red-orange poppies and a rocky outcropping. • BOTTOM RIGHT: Flowering plant of blue-eyed grass (*Sisyrinchium bellum*).

with yellow centers in spring. Grows from a few inches to about two feet tall according to form. Reseeds. Propagate by divisions or seed.

Garden Design Note: Many selections available. 'Arroyo de la Cruz' has large dark flowers; 'H-Bar-H White' looks nice interspersed with the blue form; 'Nanum' has thicker leaves and flowers the color of faded blue jeans; and 'Wayne's Dwarf' is six inches tall with dark purple flowers. Plant as an annual with poppies and tidy-tips.

Yellow-eyed grass (*Sisyrinchium californicum*).
Similar in habit to blue-eyed grass but with pale yellow flowers that bloom through spring and a second time in fall. Readily reseeds. Propagate by divisions. Prefers some summer water.
Garden Design Note: Yellow-eyed grass does better in heavy soils and is longer lived than blue-eyed grass, but treat it as an annual for best results. Likes more moisture.

Douglas iris (*Iris douglasiana*).
Bold clumps of glossy, sword-shaped leaves arranged fanwise,

TOP: Leaves and flowers of Douglas iris (*Iris douglasiana*). • BOTTOM: Douglas iris (*Iris douglasiana*) growing under a Baja birdbush (*Ornithostaphylos oppositifolia*).

237

and elegant, orchidlike flowers appearing from early to midspring. Grows to a foot high. Flower colors range from white to soft yellow, shades of lavender, purple, and blue. Many hybrids available as Pacific Coast irises, which come in a wide color range. Propagate from divisions or stratified seed. Be sure to divide clumps every few years in fall to avoid tangled roots and rhizomes.

Garden Design Note: Very versatile garden plant. Deer will not eat it. Can take full sun or light shade. 'Canyon Snow' is an outstanding selection.

Yarrow (*Achillea millefolium*).

These low, rhizomatous plants cover a large territory with sage-scented, ferny foliage that dies back in winter. Six-inch to foot-high flowering stalks carry flat-topped clusters of tiny white or pink daisies from late spring to summer, sometimes repeat blooming. Propagate by divisions. Cut back old foliage in winter. May be invasive.

Garden Design Note: Several selections produce wonderful color variations from the cream-colored species, including 'Cassis', 'Cerise Queen', 'Summer Pastels F2', and 'Paprika'. I especially like the 'Island Pink' form; it is not so invasive.

Larkspur (*Delphinium* spp.).

Summer-dormant from tubers or tough rootstocks. Buttercuplike leaves and one- to two-foot-tall racemes or spikes of lovely spurred, blue, purple, white, or pink flowers in midspring. Vulnerable to snails. *D. variegatum* (royal larkspur) has deep royal-purple flowers, but many others are also beautiful. Easily propagated from seed. Larkspurs are often short lived in the garden and should be frequently restarted from seed.

Checkerbloom (*Sidalcea malviflora*).

Close mats of palmately lobed, mallowlike leaves from short rhizomes, with spikes of rose-colored, hollyhocklike flowers in early to midspring

TOP RIGHT: Pink-flowered form of yarrow (*Achillea millefolium* 'Island Pink') at Santa Barbara Botanic Garden. Photograph by S. Ingram. • BOTTOM LEFT: Flowers of checkerbloom (*Sidalcea malviflora*).

to eight inches high. Fall/winter dormant. Propagate from seed or divide clumps in fall. Garden Design Note: Likes moisture for continuous blooming. Many bees and butterflies visit checkerbloom. Plant many in a meadow with Ithuriel's spear (*Triteleia laxa* 'Queen Fabiola').

Annual Wildflowers

Propagation Note: All annuals are propagated from seed.

Goldfields (*Lasthenia californica*).
Fragrant, miniature golden daisies carried on multibranched stems to eight inches high in midspring. Large patches of goldfields assure brilliant color for three weeks from early to midspring.

Tidy-tips (*Layia platyglossa*).
Blooms with or just after goldfields. Plants to eight inches high with large, daisylike blossoms: broad yellow ray flowers with white tips around a golden center. From a distance, patches of tidy-tips look cream colored.

Flowers of tidy-tips (*Layia platyglossa*).

Redmaids (*Calandrinia ciliata*).
Low sprawling plants with fleshy leaves and saucer-shaped, vivid magenta flowers with golden stamens. Beautiful counterpoint to blues and yellows. Blooms early to midspring.

Creamcups (*Platystemon californicus*).
Shaggy-haired plants to 12 inches high with narrow leaves, nodding pink buds, and saucer-shaped, cream-colored to yellow flowers resembling miniature poppies. Blooms midspring.

Owl's clover (*Castilleja exserta*, formerly known as *Orthocarpus purpurascens*).
Dense, moplike spikes of pink to magenta flowers on plants to eight inches high with highly divided leaves. Close inspection reveals patterns of white and yellow petals surrounded by vividly colored bracts. Beautiful combined with the blues of lupines and gilias and the gold of poppies. Blooms midspring.

Cream sacs (*Triphysaria eriantha*, formerly known as *Orthocarpus erianthus*).
Fine, feathery foliage and dense spikes of yellow blossoms trimmed with red-purple; individual petals look like little inflated balloons. Eight inches high; blooms early to midspring.

Sky lupine (*Lupinus nanus*).
Up to a foot high, with basal, palmately compound leaves and spikelike clusters of blue and white, fragrant, sweet pea-like flowers in midspring. Needs well-drained soil and full sun. Propagate from fresh seed.

Whorled lupine (*L. microcarpus* var. *densiflorus*).
Late spring-flowering species with neatly stacked whorls of sweet pea-like blossoms on shorter spikes. Flower colors include white, cream, sulfur yellow, and pale pink. An excellent selection is 'Ed Gedling', which has a rich golden color.

Globe gilia (*Gilia capitata*).
Long-blooming mid- to late-spring annual with feathery

TOP: Spikes of owl's clover (*Castilleja exserta*) flowers. • BOTTOM: The white form of whorled lupine (*Lupinus microcarpus* var. *densiflorus*) is especially dramatic.

foliage and close, globe-shaped clusters of blue to white flowers to a foot high. Combines well with tidy-tips and poppies.

Bird's-eye gilia (*Gilia tricolor*).
Low-growing species a few inches high with finely divided leaves and wide-open, saucer-shaped blossoms that combine lavender, a whitish border, and a deep purple throat spotted yellow. Blooms midspring.

Farewell-to-spring (*Clarkia amoena*).
Slender, wiry stems to a foot high carry pink or red-purple chalices sometimes spotted with red. Blooms late spring to early summer but may continue to bloom through summer with extra water. This is the most common clarkia, but several other species are available as well, including *C. speciosa*

TOP: Farewell-to-spring (*Clarkia amoena*) flowers have a red spot on each petal. • BOTTOM: Meadow with sulfur meadowfoam (*Limnanthes douglasii* var. *sulphurea*), Douglas iris (*Iris douglasiana*), and water-washed cobbles.

(showy clarkia), *C. rubicunda* (ruby chalice clarkia), *C. arcuata*, *C. bottae* (punchbowl clarkia), and *C. gracilis* (graceful clarkia). Usually reseeds.

Garden Design Note: Many cultivars of various species are available: *C. amoena* 'Fruit Punch Mix', 'Rembrandt', 'Summer Paradise', 'Thoroughly Modern Millie'; *C. purpurea* ssp. *quadrivulnera*; *C. rubicunda* ssp. *blasdalei*; *C. speciosa* ssp. *immaculata*; *C. bottae* 'Amethyst Glow' and 'Lady in Blue'; and *C. grandiflora* 'Bornita Mixed'.

Meadowfoam (*Limnanthes douglasii*).

Low, sprawling plant with pinnately divided leaves; long succession of shallow, bowl-shaped, white, yellow trimmed with white, or golden yellow flowers from early to midspring. The variety *sulphurea* has pure yellow flowers and is among the best. Often reseeds. Favors poorly drained or heavy soils.

Garden Design Note: Try *L. douglasii* var. *nivea*, whose pure white flowers appear in the spring. Does well in winter wet soil. Sow in drifts.

Baby-blue-eyes (*Nemophila menziesii*).

Early-blooming, semisprawling annual with scalloped leaves and wide-open, saucer-

shaped, clear blue flowers with a central white spot. Variety *atomaria* substitutes pale lavender petals with dark purple lines and dots. Long blooming in the garden.

Garden Design Note: 'Pennie Black' is stunning black with a white border.

Five-spot (*N. maculata*).

Similar to baby-blue-eyes except each whitish petal bears a triangle-shaped dark purple spot at its tip.

Garden Design Note: 'Chelsea Blue' is a very desirable selection available from Thompson and Morgan.

Lindley's blazing star (*Mentzelia lindleyi*).

Much-branched, mid- to late-spring-blooming plants to three feet high with stiff foliage and very large, golden, star-shaped flowers filled with wiry golden stamens. Likes heat.

Garden Design Note: Occasionally available at Larner Seeds. Difficult to establish.

Fernleaf phacelia (*Phacelia distans*).

Robust plants reaching three feet high with coarse, fernlike foliage and fiddleheads

Baby-blue-eyes (*Nemophila menziesii*).

that unfurl to shallow, bowl-shaped, blue-purple flowers with protruding stamens. Blooms midspring.

Garden Design Note: Can substitute *P. tanacetifolia* (tansy-leaf phacelia) or *P. viscida* 'Tropical Surf'; the latter has incredibly bright blue flowers that reseed annually.

Bulbs

Note: Bulbs may be started from offsets or from seed, but offspring raised from seed may be variable and usually

take three to six years to reach flowering size. Many bulbs should be lifted and stored in a dry place if the garden is watered often in summer. The following bulbs are avail-

TOP: Bunchgrass and urbanite stepping stones in an Asian themed garden. • BOTTOM: Masses of fernleaf phacelia (*Phacelia distans*) at Rancho Santa Ana Botanic Garden.

able at specialty bulb nurseries like Telos Rare Bulbs or Far West Bulbs. Most require protection from gophers and deer.

Blue dicks (*Dichelostemma capitatum*).
Flowers from March to early April with naked flowering stalks to two feet high and headlike clusters of pale blue or purple flowers surrounded by dark purple bracts. Naturalizes readily and quickly expands its area in many sites.

Golden brodiaea or pretty-face (*Triteleia ixioides*).
Stems from six to 12 inches high, with flat-topped umbels of straw-colored to golden yellow, starlike blossoms appearing in mid to late spring.

Elegant brodiaea (*Brodiaea elegans*).
Stiff stems to around eight inches high with a few funnel-shaped, waxy, blue-purple flowers of great beauty. Blooms late spring and early summer.

Soapplant (*Chlorogalum pomeridianum*).
Wavy leaves appear in winter, with flowering stalks maturing between late spring and early summer. Flower panicles may reach six feet high and provide a month-long succession of spidery white flowers that open in late afternoon and evening. Tolerates clay soils and full sun to moderate shade.

Yellow mariposa-tulip (*Calochortus luteus*).
Stiff stems to a foot high carry large chalices of golden yellow, tuliplike flowers decorated inside with lines and splotches of dark brown. Good cut flower. Blooms mid to late spring.

Venus mariposa-tulip (*C. venustus*).
Highly variable species with the same overall stature and flower shape of yellow mariposa-tulip; petals vary from pure white to lavender, pale yellow, pink, purple, and red. The varied brown and yellow markings inside mimic patterns on butterfly wings. Blooms late spring.

Additional coastal bunchgrasses to try:

California oatgrass (*Danthonia californica*), June grass (*Koeleria macrantha*), dune ryegrass (*Leymus mollis*), western fescue (*Festuca occidentalis*), California fescue (*F. californica*).

The bright yellow coastal form of pretty-face brodiaea (*Triteleia ixioides*).

TOP: Grassland wildflower meadow and turf adjacent to a flagstone path. Photograph by S. Morris. • BOTTOM LEFT: California fescue meadow (*Festuca californica*) and chaparral plantings. • BOTTOM RIGHT: Red fescue (*Festuca rubra*) meadow with creek.

Additional inland bunchgrasses to try:

Blue ryegrass (*Elymus glaucus*), California oatgrass (*Danthonia californica*), June grass (*Koeleria macrantha*), Torrey's melic grass (*Melica torreyi*), alkali sacaton (*Sporobolus airoides*).

Additional perennials to try:

California and western buttercups (*Ranunculus californicus* and *R. occidentalis*), padre's shooting stars (*Dodecatheon clevelandii*), Douglas's violet (*Viola douglasii*), blue violet (*Viola adunca*), cow parsnip (*Heracleum lanatum*), sanicles (*Sanicula* spp.).

Additional annuals to try:

Sticky and large-flowered phacelias (*Phacelia viscida* and *P. grandiflora*), whispering bells (*Emmenanthe penduliflora*), showy fiddleneck (*Amsinckia tesselata* var. *gloriosa*), thistle sage (*Salvia carduacea*), popcorn flower (*Plagiobothrys nothofulvus*), frying-pans poppy (*Eschscholzia lobbii*), Kellogg's monkeyflower (*Mimulus kelloggii*), golden monkeyflower (annual form of *M. guttatus*), glueseed (*Blennosperma nanum*), downingias (*Downingia* spp.).

Additional bulbs to try:

White brodiaea (*Triteleia hyacinthina*), goldenstars (*Bloomeria crocea*), wild hyacinth

(*Dichelostemma multiflorum*), California brodiaea (*Brodiaea californica*), dwarf brodiaeas (*B. minor* and *B. terrestris*), harvest brodiaea (*B. coronaria*), white onion (*Allium amplectens*), pink onion (*A. serra*), coast onion (*A. dichlamydeum*), paper onion (*A. unifolium*), Fremont starlily (*Zigadenus fremontii*), fragrant fritillary (*Fritillaria liliacea*), goddess mariposa-tulip (*Calochortus vestae*), superb mariposa-tulip (*C. superbus*), meadow star-tulip (*C. uniflorus*).

Middlebrook Gardens demonstration garden featuring grassland meadow and a stone bench.

Places to Visit

Coastal Prairies

Note: Most prime coastal prairie habitat lies in northern and central California.

Salt Point State Park, Sonoma County. Take Hwy 1 about 18 miles north out of Jenner. Turn left at the main entrance into the park and drive west to the ocean, where you will find informal trails; or go to the Stump Beach turnoff from Hwy 1 and hike down the quarter-mile trail to the beach and bluffs beyond. Some of the grassland is nonnative, but there are also lots of native fescues (*Festuca rubra* and *F. idahoensis*), California oatgrass (*Danthonia californica*), tufted hairgrass (*Deschampsia caespitosa*), and Nutka reed grass (*Calamagrostis nutkaensis*). Perennials include footsteps-to-spring (*Sanicula arctopoides*), coast rattlepod (*Astragalus nuttallii*), seathrift (*Armeria maritima*), California poppy (*Eschscholzia californica*), lizard-tail (*Eriophyllum stachaedifolium*), checkerbloom (*Sidalcea malviflora*), and seaside daisy

Mixed wildflowers near Gorman, Los Angeles Co.

(*Erigeron glaucus*), while among the annuals can be found tidy-tips (*Layia platyglossa*), goldfields (*Lasthenia californica*), and coast clarkia (*Clarkia davyi*). Bulbs include dwarf brodiaea (*Brodiaea terrestris*), purple pussy ears (*Calochortus tolmiei*), and soapplant (*Chlorogalum pomeridianum*).

False Lake Meadow, Annadel State Park, Sonoma County. Take Montgomery Road east from Hwy 12 at the east end of Santa Rosa, turn right onto Channel Drive, and continue to the parking lot at the far end. Follow Steve's Trail to North Burma Trail, and turn left onto Live Oak Trail, which enters the meadow. The lower meadow is a relatively unspoiled bunchgrass-land with California oatgrass and purple needle-grass (*Nassella pulchra*). Perennials include shooting stars (*Dodecatheon hendersonii*), blue-eyed grass (*Sisyrinchium bellum*), and biscuit roots (*Lomatium* spp.). Bulbs include dwarf brodiaea, white brodiaea (*Triteleia hyacinthina*), white fritillary (*Fritillaria liliacea*), yellow mariposa (*Calochortus luteus*), death-camas (*Zigadenus venenosus*), and meadow star-tulip (*Calochortus uniflorus*). Annuals include downingia (*Downingia* sp.), glueseed (*Blennosperma nan-*

um), goldfields, cream sacs (*Triphysaria eriantha*), and linanthus (*Linanthus bicolor*).

Valley Grassland

Note: Owing to the widespread destruction of most of our native grasslands, localities that display intact examples of valley grassland are widely scattered, fragmented, and difficult to find.

Bear Valley, Colusa County. Take Hwy 20 west from Williams to Bear Valley Road and drive the few miles north into the valley. This area features an open grassland with alien and scattered native grasses and magnificent wildflower displays. Bulbs include adobe-lily (*Fritillaria pluriflora*), meadow star-tulip, blue dicks (*Dichelostemma capitatum*), Ithuriel's spear (*Triteleia laxa*), white brodiaea, and mariposa-tulips (*Calochortus superbus* and *C. splendens*). Wild-flowers include tidy-tips (*Layia* spp.), California poppy, redmaids (*Calandrinia ciliata*), glueseed, goldfields, royal larkspur (*Delphinium variegatum*), creamcups (*Platystemon californicus*), and owl's clover (*Castilleja exserta*).

Hunter-Liggett area, Monterey County. From Hwy 101, take the Jolon Road exit just north of King City. After several miles, the road enters the Hunter-Liggett military reservation near the former town of Jolon. Currently, the reserve is off-limits to the perusal of wildflowers but adjacent private lands to the south are equally good. The grasslands are a combination of nonnative grasses, needlegrass, and deer grass (*Muhlenbergia rigens*) among huge valley oaks. The annual wildflower displays are among the best in the state, and include sky lupine (*Lupinus nanus*), owl's clover, redmaids, cream-cups, goldfields, baby-blue-eyes (*Nemophila menziesii*), monolopia (*Monolopia lanceolata*), foothill wallflower (*Erysimum capitatum*),

Tidy-tips (*Layia platyglossa*), Carrizo Plains, San Luis Obispo Co.

evening snow (*Linanthus dichotomus*), linanthus (*L. bicolor* and others), showy clarkia (*Clarkia speciosa*), native dandelion (*Agoseris heterophylla*), and California poppy. Bulbs include golden stars (*Bloomeria crocea*), blue dicks, Jolon brodiaea (*Brodiaea jolonensis*), soapplant, and purple amole (*Chlorogalum purpureum*).

Gorman area, Los Angeles County. Take the Gorman exit off Hwy 5 and turn right onto Gorman Post Road on the east side of town. This road skirts steep, sparsely grass-covered slopes with great sweeps of multicolored annual wildflowers including spiderling lupine (*Lupinus benthamii*), globe gilia (*Gilia capitata*), creamcups, California poppy, desert coreopsis (*Coreopsis bigelovii*), popcorn flower (*Plagiobothrys nothofulvus*), Coulter's jewel flower (*Caulanthus coulteri*), tansy-leaf phacelia (*Phacelia tanacetifolia*), and more.

Carrizo Plains, San Luis Obispo County. From Hwy 5 at Buttonwillow (San Joaquin Valley), turn west onto Hwy 58, go over the Temblor Range to the next valley, then turn south to the tiny town of California Valley. The Carrizo Plains extend south from here for around 30 miles. Besides alkali sink vegetation in the wet places, the valley is dominated by widely spaced grasses and brilliant sheets of annual wildflowers including goldfields, creamcups, baby-blue-eyes, tansy-leaf phacelia, California poppy, owl's clover, fragrant thelypodium (*Thelypodium lemmonii*), gilias, popcorn flower, and monolopia. A common perennial here is desert larkspur (*Delphinium parishii*).

Yellow bush lupine (*Lupinus arboreus*) coastal prairie, Salt Point State Park, Sonoma Co.

CHAPTER 10

CHAPARRAL: DROUGHT-ADAPTED SHRUBS FOR THE GARDEN

Spreading like a velvet tapestry over some of California's harshest landscapes, chaparral, a dense, miniature forest of mostly evergreen shrubs, is a characteristic Mediterranean-climate plant community. The shrub crowns are spaced so tightly that there is little room to pass between them, and the shrub roots so completely plumb the soil's depths for moisture that little is left to support other plants. Smaller plants are also inhibited by the toxic oils that leach from the leaves of many chaparral shrubs into the soil.

Despite the uniform appearance of this "elfin forest" from a distance, some locales support a surprisingly diverse mix of species. Each area has its own story to tell. Places with especially shallow, rocky, nutrient-poor soils, for instance, may be covered by only one or two species, such as chamise (*Adenostoma fasciculatum*) and buckbrush (*Ceanothus cuneatus*). Other areas with more diverse topography, richer soils, or that have been recently burned may support a complex mosaic of species.

Chaparral shrubs have adapted to California's hottest, driest slopes with ingenious

ways of obtaining and conserving water. Roots probe deeply, and some have nitrogen-fixing nodules to help with nutrition. The leaves are stiff and tough, often with a minimal surface area, and they may be arranged in dense clusters or be vertically oriented. Many leaves are covered as well with whitish or silvery hairs that reflect the sun's rays or trap moisture.

The bloom time in chaparral extends from midwinter, when manzanitas and silk tassel bushes are already in bloom, to June or July, when the season finishes with the flowers of yerba santa, fremontia, chamise, and toyon. Some smaller shrubs—a whole fascinating set of plants that inhabit the fringes and slopes next to the taller shrubs—bloom through summer and into fall. Among these smaller plants are monardellas, buckwheats, and hummingbird fuchsia.

LEFT: Bigberry manzanita (*Arctostaphylos glauca*) at Pinnacles National Monument. • ABOVE: Steep slopes of the Santa Ynez Mountains behind Santa Barbara are covered with chaparral.

The prevailing flower colors are blues, purples, whites, and yellows, accented by subtle pinks, hot reds, and flashes of orange.

Chaparral shrubs are designed to burn periodically—many feature fragrant leaves filled with volatile oils that, in the heat of summer and fall, evaporate and easily catch fire. Natural, lightning-caused fires have shaped the chaparral for millennia. Before our practice of fire intervention, chaparral was thought to have burned every twenty to forty years. Without fire, most chaparral shrubs become increasingly unhealthy unless they're periodically cut back.

Creating a Garden

Because of the fire hazard of flammable foliage and bark, chaparral shrubs should be planted well beyond the perimeter of buildings. Plantings that require summer water are safer to use closer in.

For an interesting chaparral-based garden design, do not ignore plants that occur at the borders of chaparral or as stages of succession after a fire. You might, for example, use large chaparral shrubs as a backdrop, and plant smaller seasonal shrubs, large bunchgrasses, vigorous perennials, and drought-tolerant ferns in front. Rather than mix many kinds of large shrubs, moreover, it's best to stick to no more than five or six kinds for a large garden, fewer for a small one. Plant shrubs with different bloom times, comple-

mentary flower colors, and interesting foliage textures, to carry the design when shrubs are out of bloom.

TOP: Chaparral garden with Idaho fescue (*Festuca idahoensis* 'Siskiyou Blue') and California fescue (*F. californica*) in flower. Featured garden photographed by S. Ingram. • BOTTOM: Flagstone entry with Idaho fescue (*F. idahoensis* 'Siskiyou Blue') and manzanita (*Arctostaphylos* 'Louis Edmonds') in featured garden.

Ideal home garden sites for chaparral plantings are south- and west-facing slopes. Chaparral plants thrive in poor soils, so amending the soil is unnecessary. If your garden has the proper aspect and is relatively flat, I recommend building undulating mounds from 12 to 14 inches tall with a fast-draining soil. Place the chaparral plants at the top of the mounds, keeping the crown high. With good drainage, hot and sunny conditions, and an appropriate watering regime, chaparral plants can live in home gardens for 25 years. (If planted in heavy clay, however, they will have a short life span of just two to five years.)

Design Notes

The southwestern exposure of this two-story ranch home at the edge of the Cupertino foothills makes it an ideal candidate for a chaparral garden. Because the house is relatively close to the street, reflective heat from the driveway, sidewalk, and road also create higher temperatures, which chaparral plants can tolerate well. Al-

Cleveland sage (*Salvia clevelandii*) in a chaparral garden (featured garden).

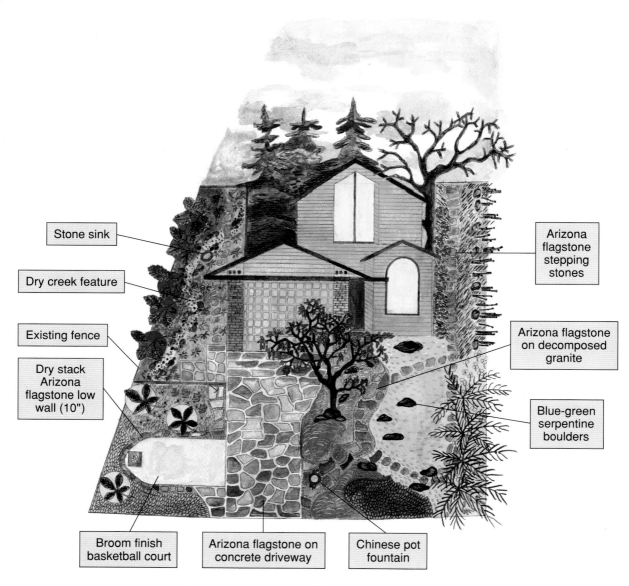

Stone sink

Dry creek feature

Existing fence

Dry stack Arizona flagstone low wall (10")

Arizona flagstone stepping stones

Arizona flagstone on decomposed granite

Blue-green serpentine boulders

Broom finish basketball court

Arizona flagstone on concrete driveway

Chinese pot fountain

ready existing on site, however, were four 20-year-old redwood trees, with a seasonal creek nearby. This situtation creates a shady microclimate at the southern edge of the garden. The planting plan and irrigation design therefore modulate from chaparral to redwood plantings very quickly.

When I first approached the house, the driveway and two-car garage dominated the view. The owners said, however, that they didn't use the right side of the garage for parking. This allowed the creation of a planting bed in front of it, as a screen. The first

Rendering of featured garden.

plant to go in was a bigberry manzanita, placed in a prominent spot. As it grows, it will add balance and decrease the visibility of the garage. At the driveway entry, a Chinese-style fountain creates soothing water noise.

On the left side of the house are three existing birch trees. Here we planted three small buckeyes, which will eventually replace the birches. As understory, we planted a small California fescue meadow. On the right side, at the property line, are the four large redwoods. Under them we placed a bench and selected redwood understory natives.

Between the fescue meadow and the redwood garden lies the chaparral. Here we chose showy species including California bush anemone, woolly blue-curls, flannelbush, silk tassel bush, and manzanita. We also added ceanothus and salvias for color and fragrance. The ground cover is kinnikinnick.

Each garden is linked by the repeated use of grasses and grasslike natives representing each community: vanilla grass from the redwoods, California fescue and Idaho fescue from chaparral and oak woodland, Berkeley sedge from the riparian, and red fescue from the cool, shaded woodland.

The most colorful garden is the front entry, which features many shades of blue, pink, oatmeal, rust, orange, yellow, and white. A buff-colored Arizona flagstone was selected for the driveway (installed on concrete) and entry (installed on decomposed granite, allowing water to percolate through). Handmade celadon Chinese stoneware containers and a stone sink add an Asian flair that is complemented by the grasses, which provide resting places for the eye as one strolls from one plant community to the next.

Two years after the garden was built, a wooden trellis was built on the garage, with California grape vines planted on each side to ramble up the framework. We're hoping this will further diminish the visual impact of the garage. The vines' red fall foliage will be a nice counterpoint to the masses of salvias at the front entry.

TOP TO BOTTOM: Manzanita (*Arctostaphylos*) • Flannelbush (*Fremontodendron*) • Chaparral clematis (*Clematis lasiantha*) • Cleveland sage (*Salvia clevelandii*). Illustrations by A. Yankellow.

Scope of Work

- ◆ Remove plants and trees.
- ◆ Remove driveway and haul away.
- ◆ Lay new driveway slab from street (where opening is 18.5 feet wide) to west two-thirds of garage. Mortar flagstone to the new slab.
- ◆ Lay decomposed granite around basketball pad and to the front door; set flagstone.
- ◆ Remove brick accent borders.
- ◆ Install flagstone in mortar on the entryway landing and at left side entry and trash holding area.
- ◆ Install flagstone paths throughout property.
- ◆ Create berms along paths to accommodate horticultural needs of certain species of plants.
- ◆ Install Arizona limestone boulders at selected spots.
- ◆ Install natural dry creek in left riparian garden.
- ◆ Install birdbath in left riparian garden.
- ◆ Install fountain at front entry.
- ◆ Install underground flexible plastic barrier to retain scouring rush in right riparian garden.
- ◆ Install creeping red fescue plugs and seeds in right riparian garden.
- ◆ Add river-washed cobbles under large trees.
- ◆ Install large flagstone pieces laid vertically against rear deck to create an elevation change.
- ◆ Install large ceramic containers with western azalea at rear deck.
- ◆ Install new irrigation system.
- ◆ Install low-voltage lights and transformer.
- ◆ Top dress planted areas with ProChip mulch.

Plant List

Chaparral

SYMBOL	BOTANICAL NAME	COMMON NAME
c	*Arctostaphylos uva-ursi*	kinnikinnick
M	*Carpenteria californica*	California bush anemone
i	*Festuca californica*	California fescue
F	*Ceanothus griseus* var. *horizontalis* 'Yankee Point'	creeping blueblossom

TOP: Tree ceanothus (*Ceanothus arboreus*). • BOTTOM: Silk tassel bush (*Garrya elliptica*). Illustrations by A. Yankellow.

K	*C. thyrsiflorus* var. *repens*	creeping wild lilac
A	*Arctostaphylos* 'Dr. Hurd'	manzanita
O	*Ceanothus* 'Dark Star'	California wild lilac
H	*Eriogonum arborescens*	Santa Cruz Island buckwheat
g	*Fremontodendron californicum* ssp. *decumbens*	prostrate flannelbush
L	*Garrya elliptica*	silk tassel bush
B	*Arctostaphylos* 'Howard McMinn'	Vine Hill manzanita
D	*Monardella odoratissima*	mountain coyote mint
j	*Festuca idahoensis* 'Siskiyou Blue'	Idaho fescue

Featured garden key.

| N | *Salvia clevelandii* | Cleveland sage |
| E | *Trichostema lanatum* | woolly blue-curls |

Riparian Garden (Right Side)

P	*Umbellularia californica*	California bay
Q	*Equisetum hyemale*	scouring rush
R	*Festuca rubra*	red fescue
S	*Iris douglasiana*	Douglas iris
T	*Juncus patens*	California rush
U	*Lilium pardalinum*	leopard lily
V	*Myrica californica*	California wax myrtle

Riparian Garden (Left Side)

W	*Alnus rubra*	red alder
X	*Carex tumulicola*	Berkeley sedge
Y	*Calycanthus occidentalis*	western spice bush
Z	*Cornus nuttallii* 'Eddie's White Wonder'	flowering dogwood
aa	*Calystegia purpurata*	western morning glory
dd	*Lilium pardalinum*	leopard lily
bb	*Physocarpus capitatus*	ninebark
cc	*Philadelphus lewisii*	western mock-orange
u	*Woodwardia fimbriata*	giant chain fern

Woodland

e	*Aesculus californica*	California buckeye
f	*Ribes sanguineum* var. *glutinosum*	pink-flowering currant
not shown	*Sisyrinchium californicum*	yellow-eyed grass

Redwood Garden

LL	*Aquilegia formosa*	red columbine
ii	*Fragaria vesca*	woodland strawberry
jj	*Hierochloe occidentalis*	vanilla grass
kk	*Oxalis oregana*	redwood sorrel

TOP: Chaparral planting in a Mediterranean style. • BOTTOM LEFT: California fuchsia (*Epilobium canum*) and California fuchsia 'Wayne's Select' (*Epilobium canum* 'Wayne's Select') in a Mediterranean garden. • BOTTOM RIGHT: Chaparral plantings with desert container gardens (from featured garden in riparian chapter).

h	*Rhododendron occidentale*	western azalea
not shown	*Sequoia sempervirens*	coast redwood
m	*Vaccinium ovatum*	evergreen huckleberry

Plants to Use

Large Shrubs

Wild lilacs (*Ceanothus* spp.).

Fast-growing shrubs with nitrogen-fixing nodules on their roots. California species range considerably in size; several reach 10 to 20 feet high, often with an equal spread. The majority have leathery, evergreen leaves of lustrous green, though some are bluish or whitish. Masses of tiny flowers, highly attractive to bees, are arranged in dense panicles or cymes of sweetly fragrant white, pink, purple, or blue. A few wild lilacs are deciduous, and a few have thorns.

Ceanothus are noted for their susceptibility to fungal diseases and in some sites are notoriously short-lived. Large shrubs include blue-blossom (*C. thyrsiflorus*), with sky-blue to white blossoms; tobacco brush (*C. velutinus*), with unusually large, curled, sticky, fragrant leaves and pure white blossoms; wart-leaf ceanothus (*C. papillosus*), with narrow, warty leaves and vivid blue flowers; red-heart (*C. spinosus*), with thorny branches and frothy clusters of pale blue-purple flowers; and thick-leaf ceanothus (*C. crassifolius*), with rigid branches, small, tough, deep green leaves, and rounded clusters of white to pale purple blossoms. There are many cultivars available in nurseries, representing hybrids and selected forms of wild species. Propagate from hardwood cuttings or stratified seed. Prune out the older branches and dead growth to open the crowns to air circulation; large specimens often have handsome bark to display. Although many species are notoriously short lived in the garden, occasional coppicing back to the roots promotes healthy new growth from the root crowns.

Garden Design Note: A selection of versatile ceanothus for a layered effect in the garden includes 'Ray Hartman' and *C. arboreus* for tall shrubs; 'Dark Star', 'Julia Phelps', 'Concha', and 'Frosty Blue' for medium height; 'Joyce Coulter', 'Joan Mirov', *C. rigidus*

No-irrigation chaparral garden. Photograph by G. Rubin.

'Snowball', and 'Arroyo de la Cruz' for higher mounding shrubs; and 'Yankee Point', *C. maritimus*, and *C. gloriosus* for ground cover. Inland, plant the crowns high and give more shade with excellent drainage.

Manzanitas (*Arctostaphylos* spp.).

Manzanitas vary enormously in size and stature, with some growing into small trees. All feature lustrous red or red-purple bark—a highlight to emphasize by appropriate pruning; stiff, elliptical to ovate leaves; and racemes or panicles of hanging, urn-shaped, white to rose-pink, fragrant blossoms from winter to early spring. Applelike berries may be mahogany colored, rosy red, or pinkish. Flowers draw many pollinators, including hummingbirds; fruits are attractive to many forms of wildlife. Some large manzanitas include whiteleaf manzanita (*A. viscida*), with whitish leaves and white to pink flowers; bigberry manzanita (*A. glauca*), with bluish to whitish leaves and white flowers in midwinter; greenleaf manzanita (*A. patula*), with clear green leaves and pink flowers; Eastwood manzanita (*A. glandulosa*), with pale green leaves, a basal burl, often sticky hairs on twigs, and white flowers; heart-leaf manzanita (*A. andersonii*), with clasping, heart-shaped, pale green leaves and white flowers; and

Bigberry manzanita (*Arctostaphylos glauca*).

several closely related species, including Alameda manzanita (*A. pallida*); King Mountain manzanita (*A. regismontana*); Montara Mountain manzanita (*A. montaraensis*); and Mt. Diablo manzanita (*A. auriculata*). As with ceanothus, there are many cultivars available in nurseries, many of which have better garden tolerance than wild varieties. Propagate from hardwood cuttings or stratified or scarified seed. Manzanitas need occasional pruning out of old dead twigs to help display the beautiful bark. They may also be tip pruned to promote bushiness, but severe pruning is not recommended, as it may cause considerable dieback or even death. Most manzanitas do not stump sprout; those with burls can be coppiced in the manner recommended for ceanothus.

Garden Design Note: Gardenworthy tall shrub forms of manzanita include bigberry

and 'Dr. Hurd'; medium shrubs are 'Louis Edmonds', 'Howard McMinn', 'Sunset', 'White Lanterns', and Pajaro; low shrub selections are 'Carmel Sur', 'John Dourley', 'Pacific Mist', and 'Canyon Sparkles'; and ground cover manzanitas are bearberry, 'Pt. Reyes', and 'Emerald Carpet' or 'Monterey Carpet'. Manzanitas require inland shade and work well under oaks. Manzanitas are slower growing than ceanothus. Plant the garden spaces between them with grasses, annuals and short-lived perennials.

Silk tassel bush (*Garrya elliptica*).
Small, dioecious, evergreen tree or shrub to 20 feet high. Pairs of leaves are leathery, wavy, dark green, broadly elliptical. The

TOP: Bigberry manzanita (*Arctostaphylos glauca*) has dramatic, blue-green leaves. • BOTTOM: Silk tassel bush (*Garrya elliptica*) is decorated with long male catkins in winter.

males have long, elegant, grayish tassels in winter; the females shorter tassels that ripen into chains of gray, grapelike fruits. Needs well-drained soil, full sun to light shade, occasional water. Propagate from hardwood cuttings or stratified seed. Cultivars include 'James Roof'. Silk tassel bushes need minimal maintenance; tip pruning may promote bushiness. Very old specimens are sometimes coppiced to the root crown to rejuvenate the plants with vigorous new growth.

Red shanks (*Adenostoma sparsifolium*).

Small, multitrunked tree to 25 feet high with a rounded crown. Long ribbons of reddish brown bark; branches covered with narrow, feathery, light green, needle-like leaves; and dense sprays of tiny white blossoms in late spring to early summer. Highly drought

tolerant and adaptable to rocky, nutrient-poor soils. From middle elevations in southern California's Transverse and Peninsular ranges. Propagate from hardwood cuttings and stratified seed. Remove old dead twigs and prune out excess branches to reveal the beautiful bark.

Garden Design Note: If you have the space, a grove of red shanks will become increasingly more spectacular with age.

Mountain mahogany (*Cercocarpus betuloides*).

TOP: Flowering branches of red shanks (*Adenostoma sparsifolium*). • BOTTOM: Flowering branches of mountain mahogany (*Cercocarpus betuloides*).

Large shrub or small tree to 20 feet high featuring stiff, upward-trending branches, pale gray bark, and elliptical, toothed leaves. Midspring blossoms are tiny yellow-green saucers borne among the leaves and followed in early summer by plumed, white fruits that glisten in the sun. Moderate growth rate. May be espaliered or pruned into various shapes. Propagate from hardwood cuttings or stratified seed. Plants often develop rather dense, tangled main stems from the base, which should be periodically thinned to improve air circulation and enhance the overall structure.
Garden Design Note: Lends itself to pruning. Excellent selection for formal hedges in gardens.

Hollyleaf cherry (*Prunus ilicifolia*).
Small tree or large shrub to over 20 feet tall with dense branches and shiny, evergreen, hollylike leaves. Close trusses of white cherry blossoms appear in midspring and are eagerly sought by bees. Large, sweet, dark red-purple fruits lure birds in late summer and early fall. Propagate from seed. Some people dislike the mess the fruits create when they drop, since they stain pavements. The seeds often volunteer, so be sure to watch for unwanted seedlings that could lead to a tangled mess later. Shear the tips to create a dense hedge.
Garden Design Note: Another great evergreen garden hedge or screening plant. Very low maintenance and virtually pest free. Sun or shade.

Fremontia or flannelbush (*Fremontodendron* hybrids).
Relatively fast-growing, evergreen shrubs or small trees to 20 feet high living 50 or more years (often shorter-lived in gardens). Rather straggly branches need pruning to stay dense. Broad maplelike leaves, stringy and shreddy bark, and copious, broad, star-shaped, yellow-orange flowers of great beauty. Flowers smother

LEFT AND RIGHT: Leaves and flowers of fremontia (*Fremontodendron* hybrid).

branches from midspring to early summer. Fremontias are excellent subjects to espalier. Most cultivars are hybrids. Low-growing forms include 'Ken Taylor', with yellow-orange flowers, and *F. californicum* var. *decumbens,* with orange or orange-pink blossoms. Propagate from hardwood cuttings (bottom heat) or stratified seed. Be sure to provide excellent drainage. Fremontia branches can be cut back considerably to improve density and avoid legginess, but be sure to use very sharp shears, as the bark easily tears. Wear gloves to protect your skin from the irritating hairs on stems and leaves. Be sure to avoid surface summer water after plants are well established; fremontias are sensitive to fungal pathogens.

Garden Design Note: One-gallon plants have a better chance of long-term survival than five-gallon plants. Flannelbush is not a good container plant, nor does it transplant easily.

Masses of cleveland sage (*Salvia clevelandii*) in a garden setting.

265

Bush anemone (*Carpenteria californica*).
Evergreen shrub to 12 feet high. Narrowly elliptical to lance-shaped, glossy leaves. Single,

large, fragrant white flowers filled with yellow stamens appear in late spring. Requires well-drained soil, little water; prefers light shade. Propagate from hardwood cuttings or seed. Prune out dead growth. Branches may also be cut back a few inches to improve density; bush anemone can make a hedge. Leaves often develop unsightly dark spots in winter due to fungal infection; if this becomes severe, remove the worst parts and burn them. Usually, the new growth outpaces the infection and the plants look good through spring and summer.

Garden Design Note: 'Elizabeth' is floriferous and more compact. Remove old brown leaves as a regular maintenance practice. With judicious pruning, bush anemone can become a very attractive flowering shrub or a small flowering standard tree.

Bush mallows (*Malacothamnus* spp.).
Attractive, fast-growing fillers that often grow aggressively. Generally narrow, multibranched shrubs from three to over 10 feet tall. Maplelike, felted leaves are pale green to silvery gray. Showy pink, purple, or white blossoms appear in profusion from midspring to summer; may rebloom in fall. Fast-growing and short lived. Propagate from suckers, cuttings, or seed. Because bush mallows grow fast, they often become leggy and should be periodically pruned to promote a bushier structure. They can also be coppiced to promote new shoots; this will help lengthen their life in gardens.

Garden Design Note: If you can find them, collect different species for a bush mallow–themed garden. 'Edgewood', Fremont's bush mallow (*M. fremontii*), and Palmer's bush mallow (*M. palmeri*) are more available during their bloom times.

Smaller Shrubs

Note: Several smaller ceanothus and manzanitas are not detailed here. Examples include *Ceanothus* 'Julia Phelps', *C. foliosus* (littleleaf ceanothus), *C. gloriosus* var. *exaltatus*, *C. cuneatus* var. *rigidus* 'Snowball', *Arctostaphylos densiflora* 'Howard McMinn', *A. nummularia* 'Small Change', *A. edmunsii* (Little Sur manzanita, coastal gardens only), *A. hookeri* var. *montana* and var. *hookeri* (coastal gardens), and *A. bakeri*.

Flowers and leaves of bush anemone (*Carpenteria californica*). Photograph by S. Ingram.

Bush poppy (*Dendromecon rigida*).
Rapid growth; spreads by underground runners to create colonies. Around six to eight feet high with stiff, bluish green, willowlike leaves and a plethora of large, saucer-shaped, clear yellow flowers of great beauty. Propagate from divisions of runners or fire-treated seed (germination will be uneven). Bush poppy often becomes leggy in age; older growth can be coppiced to the root crown to promote healthy new growth. Must have excellent drainage.
Garden Design Note: Bush poppy is sensitive to transplant, especially during summer. *D. rigida,* which blooms once each spring, is not as readily available as the Island form, *D. harfordii*, nor is the foliage as attractive.

Cleveland sage (*Salvia clevelandii*).
Small subshrub to three feet high with pairs of fragrant, dull green, elliptical leaves and spikes of clear blue, two-lipped flowers in summer. This is the last of the salvias to

Ground cover of kinnikinnick (*Arctostaphylos uva-ursi*) beneath a rustic stone wall.

bloom. Needs excellent drainage; cut back partway every year or two to maintain vigor and compactness. Propagate from semihardwood cuttings or seed.

Garden Design Note: Plant near the door or entry for its enticing aroma. Mix with other sages and penstemons. 'Whirly Blue' and 'Winnifred Gilman' are good hybrids.

Woolly blue-curls (*Trichostema lanatum*).
Short, trim shrub no more than four feet tall with bright green, narrowly lance-shaped, aromatic leaves and spike-like racemes of red-purple and blue flowers. Resembles an upright rosemary. Each blossom has woolly red-purple sepals, clear blue two-lipped petals, and curled stamens. Especially attractive to hummingbirds. Blooms from midspring to early summer and again in fall. Volunteers from seed and is easily propagated from cuttings. Needs excellent drainage. Woolly blue curls often lives only a few years, so be sure to start new plants periodically from cuttings.

Garden Design Note: This aromatic shrub will develop a gnarled, woody trunk, suitable for a bonsai effect.

Woolly yerba santa (*Eriodictyon tomentosum*).
Widely trending roots quickly establish broad-based colonies with stems to four feet high. Pleasantly aromatic, broadly elliptical, evergreen leaves are covered with silver wool. Flowers are small, pale purple urns borne in close clusters in early summer. Propagate from root divisions. Although colonies may become larger than wanted, new growth can easily be pulled out and cut away; these divisions can then serve as a means of propagating more plants.

Garden Design Note: This low shrub has beautiful gray foliage that will cover a hillside, but it is uncommon in the trade.

TOP RIGHT: Flowering plant of Cleveland sage (*Salvia clevelandii*). • LEFT: Flowering spike of woolly blue-curls (*Trichostema lanatum*).

Pitcher sage (*Lepechinia calycina*).
Fast-growing, semievergreen shrub to four feet high with softly felted, elliptical, pale green, highly fragrant leaves. Arching branches carry inflated, white to blue-purple, two-lipped, bell-shaped flowers in midspring. Fruits are surrounded by enlarged, veined sepals shaped like miniature pitchers. Propagate from cuttings or seed. Tip pruning early in the season helps promote denser plants. Plants can also be periodically coppiced to the ground to encourage healthy new growth.
Garden Design Note: Add this shrub to a sage or bush mallow garden, or plant with yellow, red, or orange-flowering perennials. 'Rocky Point' is a desirable selection.

Chaparral yucca (*Hesperoyucca whipplei*).
Huge rosettes of narrow, dagger-sharp, green to silvery leaves. After 10 to 15 years, an enormous panicle emerges, reaching eight to 12 feet high, with numerous, nodding, waxy, bell-shaped, white flowers. Plants die after setting seed but may first make a circle of "pups." Peerless contrast with other chaparral shrubs. Propagate from offsets or seed. Dead leaves and flowering stalks are highly fibrous and difficult to compost.
Garden Design Note: Be cautious with yucca. Give it ample room, and plant away from active areas.

Chaparral yucca (*Hesperoyucca whipplei*) and deerbroom lotus (*Lotus scoparius*) in fall.

Perennials

Matilija poppy (*Romneya coulteri*).
Rampant opportunist grown as a woody-based perennial. To over six feet tall, with slashed, bluish green leaves and huge terminal flowers with crumpled white petals and golden stamens. Attractive to bees. Blooms from early summer to fall. Propagate from root divisions. Be sure to allow plenty of room for the spread of the roots, which produce many new shoots each year. Cut back old stems to the ground each fall; new shoots emerge in midspring.

Garden Design Note: Plant with San Diego ceanothus on a hillside. They bloom at the same time.

Coyote mint (*Monardella* spp.).
Bushy, woody-based perennials from a few inches to a foot tall. Glossy, fuzzy or matted, oval leaves with a pleasant sage-mint aroma. Tubular blue, purple, pinkish, or near white flowers in dense heads are strong draws to butterflies and bumblebees. Bloom from late spring into summer with supplemental water. *M. villosa* is variable and of relatively easy culture; *M. purpurea* is a matted plant with smaller leaves; *M. odoratissima* creates broad colonies, its flowers ranging from palest lavender to blue-purple; *M. macrantha*—hummingbird mint—is prostrate and bears long orange to scarlet flowers attractive to hummingbirds. Coyote mints are often relatively short lived in gardens; extra summer water will prolong bloom but shorten overall life. Cuttings should be taken every few years to replace old plants.
Garden Design Note: Coyote mints are great plants for rock gardens. Mass for spectacular color. Hummingbird mint requires excellent drainage. Useful for herbs and teas.

Sonoma sage (*Salvia sonomensis*).
Prostrate, semiwoody ground cover to a foot high in flower. Fragrant, narrowly elliptical, dull green leaves and short spikes of blue-purple flowers attractive to bees. Needs perfect drainage and full sun; ideal for a slope. Resents overwatering in summer. Propagate from layered pieces, cuttings, or seed. Cultivars available.
Garden Design Note: 'Bee's Bliss' has gray foliage; 'Dara's Choice' has darker green leaves and a more upright habit. Mix all three for fast-growing, textured, aromatic ground cover.

Matilija poppy (*Romneya coulteri*).

Hummingbird fuchsia (*Epilobium canum*, and *E. septentrionale*).

Fuchsialike blossoms attractive to hummingbirds on colonizing, sub-woody perennials usually no more than three feet high. Linear to lance-shaped, gray to bright green leaves; long racemes of narrow, trumpet-shaped scarlet flowers; and capsules with numerous, hairy seeds. Bloom from August to late fall. Many are aggressive spreaders. *E. canum* has rangy stems, while *E. septentrionale* is compact. Cultivars include plants

with white or pink flowers such as 'Solidarity Pink'. Propagate from root divisions, cuttings, or seed. Be sure to cut old stems to the ground after blooms finish in fall; this promotes healthy new growth each spring.

Garden Design Note: I like 'Silver Select', 'Everett's Choice', and 'Select Mattole'.

Golden yarrow (*Eriophyllum confertiflorum*).

Shrublet from a foot to 18 inches high with gray-green, deeply pinnately lobed leaves and flat-topped clusters of golden yellow daisies attractive to bees and butterflies. Peak bloom is from mid to late spring. Propagate from seed or semihardwood cuttings. Avoid summer water.

Scarlet bugler (*Penstemon centranthifolius*).

Woody-based, upright perennial to three feet high in flower, with grayish, elliptical leaves and narrow spikes of tubular, scarlet-red flowers attractive to hummingbirds. Needs excellent drainage; tends to be short lived in gardens. Propagate from seed or semihardwood cuttings.

TOP: Flowers of hummingbird fuchsia (*Epilobium canum*). • BOTTOM: Flowering plant of scarlet bugler (*Penstemon centranthifolius*).

271

Garden Design Note: Mass these under oaks for a crimson display or mix with other penstemons and ground cover sages in the perennial border.

Blue foothill penstemon

(*Penstemon heterophyllus*).
Woody-based, bushy perennial to two feet high with narrow leaves and spikes of blue, purple, or rose-purple, two-lipped flowers of great beauty in late spring and summer. Needs excellent drainage and may be short lived with summer water. Propagate from seed or semihardwood cuttings.
Garden Design Note: 'Margarita BOP' is a sturdy selection.

Chaparral clematis

(*Clematis lasiantha*).
Woody, scrambling vine to fifteen feet high with pairs of compound leaves and clusters of starlike, creamy flowers in midspring. Use with care, as it may overwhelm small shrubs. Propagate from layering or stratified seed.
Garden Design Note: Train on a trellis to maximize its bloom and fruiting times.

Ferns

Note: Propagate all ferns from divisions or spores. All ferns have fibrous roots that need to be fully established before summer water is withheld.

Birdsfoot fern

(*Pellaea mucronata*).
Fronds are wiry-tough, three times divided into narrow, tightly

---

TOP: Blue foothill penstemon (*Penstemon heterophyllus*). • BOTTOM: Chaparral clematis (*Clematis lasiantha*) in the wild.

curled, bluish green segments. Flush of new growth in spring. The fronds remain on plants year round but look worse for wear after a long hot summer. Cut fronds back in late fall. Seeks deep rock crevices where roots remain cool. Remove old dead fronds at the end of the growing season to allow space for the new.

Garden Design Note: Tuck this fern near a boulder and give it a few years to become fully established. Provide shade.

Coffee fern (*Pellaea andromedifolia*). Fronds similar to birdsfoot fern but coarser and less divided; the leathery bluish green segments are larger. Fronds turn a deep coffee brown or reddish purple in fall. Prefers rocky

TOP: Chaparral planting with dry stacked fieldstone wall. • BOTTOM: Bush anemone (*Carpenteria californica*) and Idaho fescue (*Festuca idahoensis*) in the featured garden.

places with afternoon shade. Remove old dead fronds at the end of
the growing season to allow space
for the new.

Garden Design Note: Mix with
birdsfoot ferns. They prefer similar
garden habitats and establish slowly.

Additional large shrubs to try:

Chamise (*Adenostoma fasciculatum*), snowdrop bush (*Styrax officinalis* var. *redivivus*), common scrub
oak (*Quercus berberidifolia*), leather
oak (*Q. durata*).

Additional small shrubs to try:

Linear-leaved goldenbush (*Erica-
merica linearifolia*), golden fleece
(*E. arborescens*), flat-topped buckwheat (*Eriogonum fasciculatum*),
blue-witches (*Solanum umbelliferum* and *S. xanti*), southern yerba
santa (*Eriodictyon trichocalyx*),
white sage (*Salvia apiana*), chaparral currant (*Ribes malvaceum*),
chaparral pea (*Pickeringia montana*).

Additional perennials to try:

Woolly sunflower (*Eriophyllum lanatum*), everlastings (*Gnaphalium
californicum* and *G. microcephalum*), buckwheats (*Eriogonum*, various species), bush monkeyflowers (*Mimulus*, various species),
rushrose (*Helianthemum scoparium*).

Additional ferns to try:

Goldback fern (*Pentagramma triangularis*), Indian's dream (*Aspidotis densa*), lace fern
(*Cheilanthes gracillima*), California polypody (*Polypodium californicum*).

TOP RIGHT: Monkeyflower (*Mimulus aurantiacus*), hollyleaf cherry (*Prunus ilicifolia*), and hesperaloe (nonnative). • MIDDLE
RIGHT: California fuchsia (*Epilobium canum* 'Wayne's Select'). • BOTTOM LEFT: Matilija poppy (*Romneya coulteri*).

Chaparral garden with ledgestone walls.

Places to Visit

Cook and Green Pass, Siskiyou County. Cook and Green Pass is located a few miles north of the Klamath River out of the hamlet of Seiad Valley, almost at the Oregon border. The several plant communities include mixed conifer forest, scree, montane meadows, and a northern version of chaparral. Chaparral shrubs include greenleaf manzanita, green-leaf silk tassel bush (*Garrya fremontii*), huckleberry oak, Sadler's oak (*Quercus sadleriana*), Brewer's oak (*Q. garryana* var. *breweri*), serviceberry (*Amelanchier alnifolia*), tobacco brush, and bitter cherry. Associated herbaceous plants include northern golden brodiaea (*Triteleia crocea*), Washington lily (*Lilium washingtonianum*), pussy paws (*Calyptridium umbellatum*), and blue penstemons (*Penstemon* spp.).

Bootjack Camp, Mt. Tamalpais, Marin County. Take Panoramic Road from Mill Valley up the mountain. Bootjack Camp is located about a mile below Pantoll Ranger Station. Hike from the parking lot up to Old Stage Road to access the chaparral, or follow the Matt Davis Trail east from the lot toward Mountain Home. The chaparral here occurs on sandstone and serpentine substrates. Shrubs include mountain manzanita (*Arctostaphylos hookeri* ssp. *montana*), glossy-leaf manzanita (*A. nummularia*), Eastwood manzanita (*A. glandulosa*), chamise (*Adenostoma fasciculatum*), Jepson's ceanothus (*Ceanothus jepsonii*), buckbrush (*C. cuneatus*), little-leaf ceanothus (*C. foliosus*), golden fleece (*Ericameria arborescens*), bush poppy (*Dendromecon rigida*), chaparral pea (*Pickeringia montana*), common yerba santa (*Eriodictyon californicum*), and toyon (*Heteromeles arbutifolia*). Perennials include Indian's dream fern (*Aspidotis densa*), serpentine reed grass (*Calamagrostis ophitidis*), false lupine (*Thermopsis macrophylla*), and purple monardella (*Monardella purpurea*).

Onion Valley, east side of the Sierra, Inyo County. Take Hwy 395 south to Independence and turn west onto the road to Onion Valley, which takes you up dizzying switchbacks to a parking lot. The high montane chaparral here includes greenleaf manzanita (*Arctostaphylos patula*), mountain chinquapin (*Chrysolepis sempervirens*), huckleberry oak (*Quercus vaccinifolia*), bitter cherry (*Prunus emarginata*), tobacco brush (*Ceanothus velutinus*), snowbrush ceanothus (*C. cordulatus*), and curl-leaf mountain mahogany (*Cercocarpus ledifolius*).

Tule River canyon, southern Sierra, Tulare County. Take Hwy 190 east from Porterville past Springville into the Tule River canyon. The middle elevations feature extensive chaparral interspersed with oak woodland and grassland; yellow pine forest lies above at higher elevations. Shrubs include fremontia, bush poppy,

Bigberry manzanita (*Arctostaphylos glauca*) has dramatic, blue-green leaves.

common yerba santa, whiteleaf manzanita (*Arctostaphylos viscida*), white-thorn (*Ceanothus leucodermis*), buckbrush, chaparral yucca, chamise, Sierra plum (*Prunus subcordata*), and the rare bladdernut (*Staphylea bolanderi*). Plants of note for the area include rose globe-tulip (*Calochortus amoenus*), Munz's iris (*Iris munzii*), twining brodiaea (*Dichelostemma volubile*), hot-rock dudleya (*Dudleya cymosa*), blue foothill penstemon, and golden brodiaea (*Triteleia ixioides*).

Camino Cielo, Santa Ynez Mountains, Santa Barbara County. From Hwy 152 at the top of San Marcos Pass, turn south (east) onto East Camino Cielo. This road follows the spine of the Santa Ynez Mountains to La Cumbre Peak. Much of the area is chaparral interspersed with coastal sage scrub, oak woodland, and Coulter pines (*Pinus coulteri*). Shrubs include bush poppy, chaparral pea, common and woolly yerba santas (*Eriodictyon trichocalyx* and *E. tomentosum*), thickleaf ceanothus (*Ceanothus crassifolius*), red-heart ceanothus (*C. spinosus*), bigpod ceanothus (*C. megacarpus*), chamise, toyon, fuchsia-flowered gooseberry (*Ribes speciosum*), snowdrop bush (*Styrax officinalis* var. *redivivus*), woolly blue-curls (*Trichostema lanatum*), chaparral yucca (*Hesperoyucca whipplei*), scrub oak (*Quercus berberidifolia*), and white and black sages (*Salvia apiana* and *S. mellifera*). Herbaceous plants include scarlet larkspur (*Delphinium cardinale*), golden and creamy eardrops (*Dicentra chrysantha* and *D. ochroleuca*), blue foothill penstemon (*Penstemon heterophyllus*), deer-broom lotus (*Lotus scoparius*), and California peony (*Paeonia californica*).

Laguna Mountains, San Diego County. From the town of Julian, take Hwy 79 south and turn left onto Hwy S-1 near Cuyamaca Lake. The road passes several dramatic overlooks with views down precipitous canyons to the desert

badlands to the east. The vegetation is a patchwork of mixed ponderosa pine–black oak forest and chaparral. The chaparral features holly-leaf cherry (*Prunus ilicifolia*), sugarbush (*Rhus ovata*), Parish's blue curls (*Trichostema parishii*), white-thorn ceanothus, manzanitas (*Arctostaphylos* spp.), deerbroom lotus (*Lotus scoparius*), white sage (*Salvia apiana*), hollyleaf redberry (*Rhamnus ilicifolia*), chaparral yucca, chamise, southern yerba santa (*Eriodictyon trichocalyx*), narrowleaf goldenbush (*Ericameria linearifolia*), Muller's scrub oak (*Quercus cornelius-mulleri*), and scarlet bugler penstemon (*Penstemon centranthifolius*).

Deerbrush (*Ceanothus integerrimus*) and green-leaf silk tassel bush (*Garrya fremontii*), Cook and Green Pass, Siskiyou Co.

CHAPTER 11

RIPARIAN WOODLAND: A PLANT PALETTE FOR HEAVY SOILS

Year-round streams and rivers are fringed with a classic deciduous hardwood forest of soaring trees, vigorous shrubs, and exuberant vines, all growing at a rapid pace because of constant, dependable water. Our riparian woodlands more closely evoke the sense of eastern hardwood forests than any other California plant community. The canopy of trees features massive cottonwoods (*Populus* spp.), muscular sycamores (*Platanus racemosa*), broad-canopied maples (*Acer* spp.), and slender alders (*Alnus* spp.). Beneath this emergent canopy, which may reach from 50 to 100 feet high, grow small trees and large shrubs—willows and ashes (*Salix* and *Fraxinus* spp.), water birch (*Betula occidentalis*), elderberries (*Sambucus* spp.), and flowering dogwood (*Cornus nuttallii*). Draped over this understory, sometimes creating impenetrable green curtains, are seasonal vines such as clematis (*Clematis ligusticifolia*), wild morning glory (*Calystegia* spp.), vine honeysuckles (*Lonicera* spp.), California blackberry (*Rubus ursinus*), and wild grape (*Vitis californica*). To complete the picture, dense colonial undershrubs—snowberry (*Symphoricarpos albus*), California wild rose (*Rosa californica*), ninebark (*Physocarpus capitatus*), and twinberry honeysuckle (*Lonicera involucrata*)—compete for light. On sandy shores, seasonal herbaceous plants—horsetails

LEFT: Flowering branches of Pacific mountain dogwood (*Cornus nuttallii*). • TOP TO BOTTOM: Riparian plant communities.

TOP: Featured garden with grassland and riparian themes. • BOTTOM LEFT: Riparian theme including an historic cast iron fountain and container gardens (featured garden). • BOTTOM RIGHT: Detail of California fescue (*Festuca californica*) meadow with container garden accent (featured garden).

and scouring rushes (*Equisetum* spp.), mugwort (*Artemisia douglasiana*), and sedges (*Carex* spp.)—take the opportunity to colonize. Anchored between streamside boulders are tough, rhizomatous plants such as Indian rhubarb (*Darmera peltata*), rushes (*Juncus* spp.), brook orchid (*Epipactis gigantea*), and leopard lily (*Lilium pardalinum*).

Most of the woody plants here—trees, shrubs, and vines—are winter deciduous, and there's considerable seasonal drama as a result. Many feature alluring bark, intriguing architecture, dramatic fruits, and leaves that turn color in fall.

Only a few species are present at any one locale; in fact, the composition of riparian forests varies greatly according to whether there's cool summer fog, hot summer sun, or cold snowy winters. Typical forests have two or three species of emergent trees, a couple of kinds of understory trees, a few vines, and several shrubs and herbaceous plants.

Creating a Garden

For the most realistic garden composition, choose plants that occur naturally together. A riparian area is ideal, of course, if you're lucky enough to have a stream on your property. Dry or wet stream courses can be constructed, but consider that riparian plantings require water year round. Riparian species are also appropriate to sites with heavy, poorly drained soils or next to watered areas such as lawns.

To emulate nature, choose at least one emergent tree, a large understory shrub or small tree, a few lower-growing shrubs, a vine or two, and a handful of herbaceous species.

Design Notes

This Santa Clara Valley home is located on a small bluff about 200 yards from Coyote Creek, overlooking the creek and a neighboring park that is punctuated with mature live oaks. The neighborhood is historic—most of the homes were built prior to 1926—and reflects the diverse architectural styles of that period. This particular house, an architectual jewel, is built in the Spanish eclectic style.

Most of the houses on the street have expansive green lawns. These owners, however, wanted to lose the lawn and recycle their mower. Because of the close proximity of the creek, a riparian plant palette was a solid choice for a lawn substitute that would

Detail of curb strip with riparian theme (featured garden).

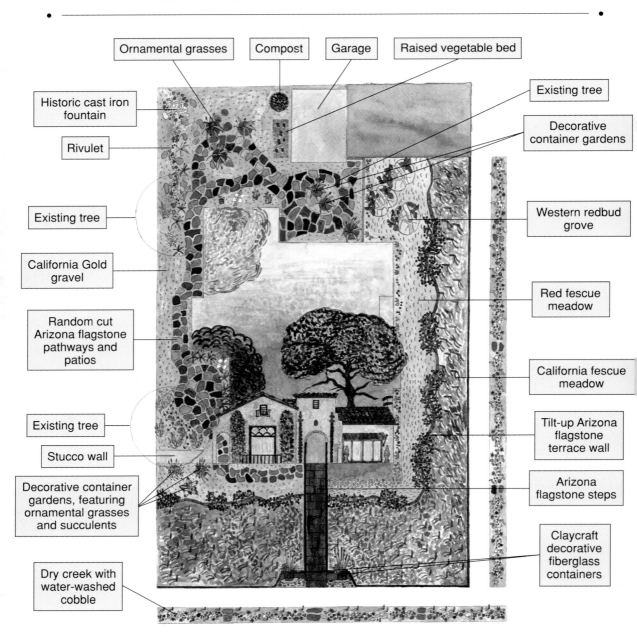

Ornamental grasses

Compost

Garage

Raised vegetable bed

Existing tree

Historic cast iron fountain

Decorative container gardens

Rivulet

Western redbud grove

Existing tree

California Gold gravel

Red fescue meadow

Random cut Arizona flagstone pathways and patios

California fescue meadow

Existing tree

Tilt-up Arizona flagstone terrace wall

Stucco wall

Decorative container gardens, featuring ornamental grasses and succulents

Arizona flagstone steps

Claycraft decorative fiberglass containers

Dry creek with water-washed cobble

be uniform in texture and color, complement the house, and fit in to this well-tended neighborhood. The new plants would be woven in to embrace a number of trees already existing on the site, including large deciduous maple, mulberry, and persimmon trees.

Because this garden has a riparian theme, stone, rock, and cobbles figure promi-

Rendering of featured garden.

nently throughout. The curbside planting strips were transformed into meandering arroyos with a mixture of water-washed cobbles in assorted colors and sizes and California fescue and Idaho fescue 'Siskiyou Blue' planted along its edges.

In the rear garden, we created a meandering dry creek that terminates at a colonial cast iron fountain that contains common scouring rush growing in a clay pot. The dry creek is

bordered with Berkeley sedge, leopard lilies, giant chain ferns, and foothill meadow-rue.

The plane of the former lawn area is broken by a terrace that is retained by vertically tipped flagstone. A lower terrace is planted with California fescue plugs; though commonly found under oaks, California fescue also thrives on summer water and is compatible with a riparian habitat. On the lower terrace, too, we replaced a liquidambar tree with a coast live oak, and a bubbler was installed to provide deep watering for the young tree. (When establishing a young oak tree, it is important to provide copious amounts of water at weekly intervals, rather than small amounts daily or several times a week.)

The upper terrace is seeded with red fescue. To maintain its rich green color, this grass requires some summer water. If summer water is withheld, it will turn yellow but will revive with winter rains. At the edge of the upper terraces chaparral and coastal bluff species, including various shades of blue and white spring-blooming ceanothus, drape over the flagstone wall.

Decorative containers are randomly placed along pathways throughout the garden. They are filled with exotic ornamental grasses and dramatic succulent forms both native and nonnative.

The owner's fondness for ornamental grasses persuaded us to select giant rye grass 'Canyon Prince' to

TOP: Riparian element with a pot fountain in urbanite patio. • BOTTOM: Sycamore (*Platanus racemosa*) and peregrine falcon in mosaic by artist Christina Yaconelli.

accent another arroyo that borders the house next to the red fescue meadow. In areas dominated by star jasmine we planted ninebark and California grape to eventually take over the fragrant exotic.

This is a simple garden plan with lots of open areas for strolling and conversation, and tree-mounted overhead lighting. It has nice water sound, and two constructed vegetable beds provide fresh vegetables and herbs to the kitchen. Although the upper and lower meadows require weed whacking one or two times a year, depending on the owners' taste for blooming grasses, it's doubtful they miss the lawn mower.

In the two years after this garden was built, a few problems had to be dealt with: one was the continued occurrence of Bermuda grass from the former lawn; the second was the ongoing loss of several of the chaparral species on the upper terrace. In the latter case, we found a solution by adjusting the watering schedule for the thirsty red fescue meadow and the drought-loving chaparral plants. Once the establishment period of two years was past, the watering schedule could be cut back and the plant losses stabilized.

As for the Bermuda grass, the key lies in preliminary preparation. If a site has established Bermuda grass, the area should be covered with plastic for at least three months during the summer heat, with no water provided to the site. It is much easier then to establish a bunchgrass meadow, without competition from the Bermuda grass.

Scope of Work

- Demolish existing plants. Rough-grade terrace area. Tie downspouts to landscaped areas with flexible drain pipe.
- Resurface existing retaining wall, existing steps, front entry, and driveway to match color of Arizona flagstone Sedona Red.
- Redesign front entry at street with wider steps and planting pockets with accent lighting at either side.
- Add water-washed cobbles and two Arizona limestone boulders at either side of the entry steps, where the flagstone terraced wall is located.
- Retrofit irrigation to accommodate new plantings.
- Install flagstone walkways and patio.
- Install California fescue plugs and seed red fescue meadow.
- Install dry creek bed in curbside planters, securing permanently with concrete. Install flagstone at curbside.
- Install curved flagstone landscape wall. Install bender board behind chaparral plantings.
- Install water-washed cobble between flagstone at pathway leading to left patio entry.

- ◆ Install bender board on right side of house between tall decorative grasses and red fescue meadow.
- ◆ As mulch, install iron oxide-tinted ProChip (to complement earth tones of house and match red tile roof) four inches deep.
- ◆ Install California Gold gravel at selected locations.
- ◆ Supply water and power for future pond construction.
- ◆ Install low-voltage path and overhead lights.
- ◆ Install raised trellis at back property line. Plant wild grape and clematis to climb trellis.

CLOCKWISE FROM TOP LEFT: California rush (*Juncus patens*) • Meadow-rue (*Thalictrum fendleri*) • California grape (*Vitis californica*) • Mock orange (*Philadelphus lewisii*) • Spice bush (*Calycanthus occidentalis*) • Wild rose (*Rosa californica*) • Sycamore (*Platanus racemosa*). Illustrations by A. Yankellow.

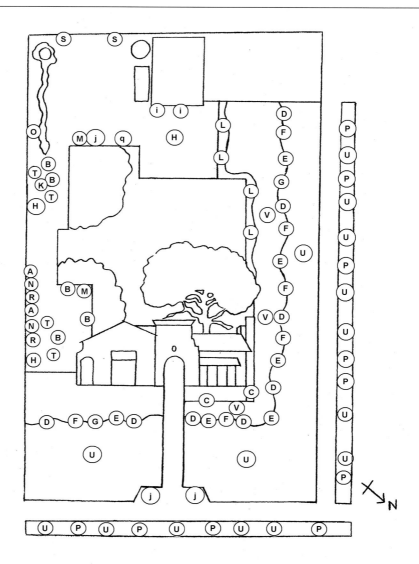

Plant List

SYMBOL	BOTANICAL NAME	COMMON NAME
A	*Aquilegia formosa*	red columbine
B	*Carex tumulicola*	Berkeley sedge
C	*Cercocarpus betuloides*	mountain mahogany
D	*Ceanothus gloriosus* 'Pt. Reyes'	glorymat
E	*C.* 'Joan Mirov'	California lilac
F	*C.* 'Julia Phelps'	California lilac

Featured garden key.

G	*C. rigidus* 'Snowball'	California lilac
H	Existing plant or tree to remain	
i	*Iris douglasiana*	Douglas iris
j	*Juncus patens*	California rush
K	*Lilium pardalinum*	leopard lily
L	*Leymus condensatus* 'Canyon Prince'	giant rye grass
M	*Philadelphus lewisii*	western mock-orange
N	*Physocarpus capitatus*	ninebark
O	*Rosa californica*	California rose
P	*Festuca idahoensis* 'Siskiyou Blue'	Idaho fescue
q	*Calycanthus occidentalis*	western spice bush
R	*Thalictrum fendleri* var. *polycarpum*	meadow-rue
S	*Vitis californica* 'Roger's Red'	California grape
T	*Woodwardia fimbriata*	giant chain fern
U	*Festuca californica*	California fescue
V	*F. rubra*	red fescue

Plants to Use

Tall Trees

Western sycamore (*Platanus racemosa*).
Bold, multiple-trunked tree to 60 feet high with jigsaw puzzle-like bark in tan, white, and gray. Large, soft-textured, maplelike leaves with collar-shaped stipules. Male and female flowers are in ball-like heads on drooping panicles, both on the same tree. Rough, ball-shaped seed heads break apart when ripe. Propagate from stratified seed. Prune out suckers at the base of the trunk. Watch for athracnose infections, which cause a premature browning of the leaves. Trees can be messy.
Garden Design Note: Plant a sycamore where you have plenty of room and adequate summer water. Loosely stake a young tree to let it develop a natural trunk. Can grow almost horizontally, or one or two can be trained as a living garden arbor.

Bigleaf maple (*Acer macrophyllum*).
Broad-canopied tree to 50 feet high, in age developing immense side branches. Large, apple-green, five-lobed leaves are coppery red when new. Hanging catkins of pale yellow flowers emerge with new leaves and attract bees. Pendant chains of doubly winged samaras ripen in fall, and in late fall leaves turn gold. Propagate from stratified or scarified seed.

Garden Design Note: Provides wonderful fall color with wild grape 'Roger's Red', *Ribes* 'King Edward VII', and flowering dogwood 'Eddie's White Wonder'.

White and red alders (*Alnus rhombifolia* and *A. rubra*).
Slender, fast-growing, relatively short-lived, many-branched trees to 50 feet high with nitrogen-fixing nodules on their roots. Roots have a bright red bark. Simple, broadly elliptical leaves have a doubly serrated edge. Slender male catkins appear in late winter to early spring, followed by shorter, upright female catkins on the same tree. Female catkins resemble miniature cones much like redwood seed cones. Propagate from seed. Prune out unwanted suckers at the base of the trunks.
Garden Design Note: Fast-growing trees that like summer water. Can take sun or shade. Will help stabilize a bank.

Young white alders (*Alnus rhombifolia*) at Fall River.

Shorter Trees

Mountain dogwood (*Cornus nuttallii*).
Slender trees to 20 feet high with tiers of branches and broadly elliptical leaves appearing in midspring. Saucerlike clusters of glorious white flowers—in reality pincushionlike heads of greenish flowers surrounded by fleshy white petal-like bracts. Red-orange, multiple fruits attract wildlife in late summer. Leaves turn orange, rose-purple, or pink in October. Propagate from layered branches, suckers, or stratified seed. Also try plants grafted to the rootstock of the eastern *C. florida,* such as 'Eddie's White Wonder'.

Water birch (*Betula occidentalis*).
Deciduous, multitrunked tree to 30 feet tall with a broad, rounded crown and drooping side branches. Small, ovate, serrated leaves turn yellow in fall. Inconspicuous greenish male and female catkins appear in early spring. Accepts many soils, full sun to light shade, summer water. Propagate from seed. Since these trees are multitrunked, you may want to open up their structure by pruning out several of the main canes.

Quaking aspen (*Populus tremuloides*).
Multitrunked, short trees with rounded crown and white to pale greenish bark. Broad, nearly round, toothed leaves turn to gold, orange, or yellow in fall with chilly nights. (Leaves fail to turn color in foothill gardens.) Flowers, which appear seldom, are borne in hanging catkins, male and female on separate trees. Propagate from suckers, or seed when available. Prune out the often vigorous suckers that would otherwise create an expanding population of crowded new trees. Best in northern California.

Willows (*Salix* spp.).
Usually multitrunked trees or shrubs, depending on species and pruning practices. Broad, rounded crowns of lance-shaped leaves, smooth margined or toothed, dark green to silvery. Twigs highly flexible. Upright catkins of male or female flowers appear on separate trees in early spring; female catkins ripen with white, cottony seeds that create a mess. Leaves turn fallow gold with autumn chill, and winter twigs are often yellow, orange, or red. There are many species to choose from. Propagate from cuttings. Willows can be severely pruned or coppiced to shape them and keep them within reasonable bounds. Be sure to watch for unwanted seedlings, which otherwise create a dense and unhealthy growth. Bunches of willow twigs bundled together can be buried on steep banks to hold soil in place.
Garden Design Note: Use fast-growing willows to create garden structures like arbors and pergolas. Train young trees to grow horizontally on wires. Prune frequently to manage new growth when young.

Shrubs

California wild rose (*Rosa californica*).
Colonizing, deciduous shrub to six feet high. Typical pin-nately compound leaves; stems lined with stout prickles. Flat-topped clusters of fragrant, pink, single-rose flowers appear throughout the summer. Bright red-orange hips at-tract wildlife in fall. Propagate from root divisions. Be sure to wear gloves when you cut canes to the ground in late fall.
Garden Design Note: Excellent plant in the habitat gar-den. Takes management.

Western spice bush (*Calycanthus occidentalis*).
Deciduous, multibranched shrub to 10 feet tall. Spicily fragrant bark and broad, ovate, paired leaves. Single, maroon-red, waterlily-shaped flowers in summer with the fragrance of old wine. Decorative seed "cups" remain through winter.

Propagate from layered branches or cuttings. Plants can be sheared into a dense hedge.
Garden Design Note: A good container plant. Large green, roundish leaves look similar to cit-rus. Flower fragrance is a novelty. Plant near a water feature.

Twinberry honeysuckle
(*Lonicera involucrata*).
Deciduous, multibranched shrub to eight feet tall. Pairs of ovate leaves. Spikes of tubular orange flowers inside pairs of red bracts in spring, with black berries rip-

TOP RIGHT: Branches of California wild rose (*Rosa californica*) with ripe hips. • MIDDLE RIGHT: Leaves and flowers of western spice bush (*Calycanthus occidentalis*). • BOTTOM: Twinberry honeysuckle (*Lonicera involucrata*) in the featured garden.

ening inside bracts and attracting birds. Propagate from layered branches or cuttings. Tip pruning often improves the density of growth. If a less dense bush is wanted, you can remove many of the major branches to open up the structure.

Garden Design Note: You can use this plant like a vine. It has an open structure. Branches are interesting as they twine though a vertical trellis.

Ninebark (*Physocarpus capitatus*).
Deciduous, thicket-forming shrub up to 10 feet tall with many branches from the base, often growing lank without pruning. Palmately lobed leaves superficially resemble currant leaves. Round heads of white, apple blossom-type flowers appear in late spring, followed by bright

red, inflated seed pods. Propagate from suckers, layered branches, or cuttings. Ninebark can be tip pruned to improve density or coppiced all the way to the ground every few years to encourage healthy new growth.

Garden Design Note: Very attractive in flower or for fall color. Mass ninebark to provide cover for birds in a habitat garden with riparian theme.

Western mock-orange
(*Philadelphus lewisii*).
Deciduous shrub with widely arching branches; needs frequent shaping. Pairs of broadly elliptical, coarsely toothed leaves. Dense racemes of white, roselike flowers with orange-blossom fragrance appear in late spring. Propagate from suckers, layered branches, or cuttings.

Garden Design Note: Very showy shrub. 'Goose Creek' has large, double flowers; 'Covelo' has two-inch flowers.

TOP: Leaves and flowers of ninebark (*Physocarpus capitatus*). • BOTTOM: Flowers of western mock-orange (*Philadelphus lewisii*).

Red-twig or creek dogwood (*Cornus sericea*).

Deciduous, suckering, fast-growing shrub to 12 feet high with bright red new twigs and pairs of ovate leaves that turn color in fall. Cymes of tiny, star-like white flowers are sporadic in spring and summer. Prune frequently to promote new growth with red twigs. Propagate from suckers, layering, or cuttings.

Garden Design Note: Excellent in containers for winter color.

Golden currant (*Ribes aureum* and varieties).

Deciduous, suckering, colonizing shrub to eight feet high with small, glossy, palmately lobed leaves. Clusters of bright yellow, sometimes fragrant flowers appear in early to midspring; ed-

ible red berries follow. Fast growing and trainable into different forms; makes an excellent espalier. Cut out unwanted suckers and allow plenty of space for plants to spread. Propagate from cuttings, layering, or divisions.

Garden Design Note: Prolific producer of sweet-

LEFT: Bare winter twigs of red-twig dogwood (*Cornus sericea*). • BOTTOM RIGHT: Golden currant (*Ribes aureum*) espalier.

tasting, juicy berries good for jellies and desserts. Hybrids of *R. sanguineum* x *R. aureum* are available from Oregon nurseries.

Vines

Western and coast morning glories (*Calystegia occidentalis* and *C. purpurata*). Winter deciduous, with herbaceous growth dying back to woody stems; growth may reach 10 to 15 feet long. Arrowhead-shaped leaves and large, single, flared white flowers, often flushed pink on the outside. Blooms over a long period beginning in midspring. Noninvasive roots. Propagate from scarified or soaked seed. Cut old herbaceous growth back to the woody stems to promote new growth next year.
Garden Design Note: The pink coastal form and the 'Anacapa Pink' selection are very desirable; both will cover a fence in one season. Sometimes there is considerable dieback, in which case new replacement plants can be planted.

Virgin's bower (*Clematis ligusticifolia*). Winter deciduous to woody stems; growth reaches 10 to 15 feet long; may overwhelm branches of trees and shrubs. Pairs of pinnately compound leaves, with dense racemes of creamy, starlike flowers appearing in early summer. Late-summer fruits are white plumed puffs as decorative as flowers. Propagate from suckers or seed. Since clematis has the potential to become invasive, the newer growth can be sheared back every year or two to the old woody stems.

California pipevine (*Aristolochia californica*). Winter deciduous to near base of plant; older plants develop woody stems. Soft, heart-shaped, pale green leaves emerge after the curious, brownish, pipe-shaped blossoms have opened in late winter. Flowers attract and temporarily trap tiny midges and beetles. Leaves provide food for the larvae of the beautiful pipevine swallowtail butterfly. Fluted seed pods hang from branches in late spring. Propagate from layered stems or cuttings. Use as a ground cover where there are no shrubs to climb.

California grape (*Vitis californica* 'Roger's Red') in the fall.

Garden Design Note: Slow growing and sometimes difficult to establish as a vigorous vine. Excellent in a children's garden.

California grape (*Vitis californica*).
Winter deciduous to woody stems, which become ropelike in age. Vines grow to 20 or more feet long. Nearly round, shallowly lobed, bright green leaves turn vivid shades of yellow or red in late fall; try the cultivars 'Roger's Red' and 'Russian River' for reliable color. Pendant panicles of tiny, yellow-green flowers in early summer attract bees. Purple, seedy grapes in fall. Propagate from cuttings or suckers. California grape can be invasive; be sure not to grow it on small shrubs and trees where it can overwhelm the crowns. Cut back newer growth in late fall to the old wood to promote new shoots next year.

California grape (*Vitis californica*) growing on a redwood arbor in a grassland/wildflower meadow in a cottage-style garden.

Garden Design Note: Good selection to cover a fence in one or two seasons. Train on a trellis for sun protection or vine tracery.

Grasslike Plants

Torrent or stream sedge (*Carex nudata*). Winter-dormant, clump-forming sedge to two feet high with gracefully drooping, grasslike leaves. White and yellow flower spikes, male above female, in early spring are wind pollinated. Propagate from divisions or seed. Old colonies can be divided to promote new growth; often the oldest clumps die out.

Garden Design Note: An excellent grasslike plant, large and bright green, for dry creeks and near water features.

TOP: Planting of torrent sedge (*Carex nudata*) under a buttonwillow (*Cephalanthus occidentalis*). • BOTTOM RIGHT: Torrent sedge (*C. nudata*).

Rushes (*Juncus* spp.).

Rather formal-looking, stiff bunches of stems bear small umbels of bronze, starlike flowers near their tips; leaves are reduced to scales in most species. Typical species for the garden include *J. effusus*, *J. acutus* (which may reach four or five feet tall), and *J. patens*, the latter with bluish green foliage. Other rushes are seldom used. Propagate from divisions or seed. Rushes should be divided every two or three years; otherwise the old clumps tend to die out and are difficult to extricate from the new, healthy growth.

Garden Design Note: Excellent in containers, providing dramatic punctuations in modern or contemporary gardens.

Ferns and Fern Relatives

Propagation Note: All ferns and their relatives are propagated by divisions or spores.

Giant chain fern (*Woodwardia fimbriata*).

Large, winter-dormant (in cold areas) fern to six or eight feet tall with coarsely twice divided fronds and linear, chainlike groups of sori on the back that show up as "stitch-

Giant chain fern (*Woodwardia fimbriata*) under sycamores (*Platanus racemosa*) in a natural creek bed.

ing" on the upperside. Remove old fronds at the end of each year.

Garden Design Note: This bold fern, California's largest, is excellent for large-scale plantings and can be massed under large shade trees for dramatic effect. Likes water.

Common horsetail (*Equisetum arvense*)

Not a true fern. Stalks to three feet high with feathery whorls of green side branches and blackish, spore-bearing cones borne on separate white stalks in early spring. Plants have invasive, wandering rhizomes and must be contained. Cut off and compost the old shoots in winter.

Garden Design Note: If left unchecked, this plant will roam. Its linear form lends itself to minimalist styles, such as Asian and contemporary.

Common scouring rush (*Equisetum hyemale*).

Slender, jointed, bamboolike, dark green stems to six feet tall with whorls of blackish, scalelike leaves. The rhizomes are invasive, so plants must be contained.

Garden Design Note: This rush is also appropriate with contemporary architecture. Use in containers.

Herbaceous Perennials

Indian rhubarb or umbrella plant (*Darmera peltata*).

Giant, winter-dormant perennial herb. Two- to three-foot tall flowering stalks in spring bear pretty clusters of pale pink flowers that ripen into red seed pods. Leaves unfurl with or after flowers into lobed umbrellas two feet across. Leaves turn bronze in fall. Propagate from divisions of rhizomes. Rhizomes tend to become tangled after a few years; cut apart and separate them with a sharp knife in winter.

Brook orchid (*Epipactis gigantea*).

Winter-dormant, rhizomatous plant that creates large, closely knit colonies. Stems with lance-shaped leaves emerge in midspring, followed in early summer by racemes

of beautiful, curious orchidlike flowers in combinations of brown, pink, green, and yellow. Propagate from divisions of rhizomes.

Garden Design Note: Plant this orchid near a water feature with seating nearby. Flowers are small. It's good with five-fingered fern and seep monkeyflower.

Leopard lily (*Lilium pardalinum*).

Winter-dormant, scaly bulbs that creep to es-

Leopard lily (*Lilium pardalinum*).

tablish large colonies. Three- to four-foot high stems bear whorls of glossy, narrow leaves and one to many large, nodding orange flowers. Each flower has six recurved petals with dark spots, and long stamens bearing rust-colored pollen. Attracts butterflies. Propagate from divisions of bulbs or seed. Colonies of bulbs often become crowded and bloom less after a few years; lift and divide these in winter.

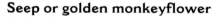

Scarlet monkeyflower (*Mimulus cardinalis*).
Clump-forming, winter-dormant perennial; short lived in the garden. Leafy stalks to three feet high with sticky, narrowly ovate, toothed leaves and clusters of large, scarlet, two-lipped flowers attractive to hummingbirds in summer and fall. 'Santa Cruz Island Gold' has golden orange flowers. Propagate from divisions or seed. Often volunteer from seed in the garden. Remove old flowering stalks to promote new ones, and cut back the old stems to the roots in late fall.

Garden Design Note: Plant beside seep monkeyflower.

Seep or golden monkeyflower (*Mimulus guttatus*).
Small clumps of leafy stems to three feet high with rounded, toothed leaves and racemes of vivid, golden yellow, two-lipped flowers sprinkled with brown spots. Blooms from midspring to midsummer. Propagate from division or seed. Short lived in the garden.

Brook saxifrage (*Boykinia major*).
Winter-dormant perennial. Basal clumps of large, round leaves are deeply palmately lobed. Foot-high flowering stalks carry flat-topped clusters of white, bergenialike flowers in summer. Propagate from divisions or seed.

Additional tall trees to try:

Black cottonwood (*Populus balsamifera* var. *trichocarpa*), Fremont cottonwood (*P. fremontii*), Oregon ash (*Fraxinus latifolia*), box elder (*Acer negundo* var. *californicum*), California bay (*Umbellularia californica*).

TOP RIGHT: Scarlet monkeyflower (*Mimulus cardinalis*). • BOTTOM LEFT: Seep monkeyflower (*M. guttatus*).

Additional shorter trees to try:

Mountain alder (*Alnus incana* var. *tenuifolia*), black hawthorn (*Crataegus suksdorfii*), native crabapple (*Malus fusca*), western yew (*Taxus brevifolia*).

Additional shrubs to try:

Black-fruited dogwood (*Cornus sessilis*), red elderberry (*Sambucus racemosa*), twin-berry honeysuckle (*Lonicera conjugialis*), stink currant (*Ribes bracteosum*), swamp goose-berry (*R. lacustre*).

Additional grasslike plants to try:

Berkeley sedge (*Carex tumulicola*), deer grass (*Muhlenbergia rigens*), alkali sacaton (*Sporobolus airoides*), giant reed grass (*Phragmites australis*).

Additional perennials to try:

Blue lobelia (*Lobelia dunnii*), scarlet lobelia (*L. cardinalis*), northern larkspur (*Delphinium trolliifolium*), red columbine (*Aquilegia formosa*), musk monkeyflower (*Mimulus moschatus*), lemon lily (*Lilium parryi*), seep goldenrod (*Solidago confinis*), western goldenrod (*Euthamia occidentalis*).

Additional ferns to try:

Lady fern (*Athyrium filix-femina*), deer fern (*Blechnum spicant*), Venus hair fern (*Adiantum capillus-veneris*), Sierra water fern (*Thelypteris nevadensis*).

Constructed riparian feature.

Places to Visit
Coastal Watercourses

Eel River near the Avenue of the Giants, Humboldt County. Just off Hwy 101 north of Garberville is access to the broad floodplain of the Eel River, where riparian corridors are backed by giant coast redwoods. Dominant trees include red alder (*Alnus rubra*), bigleaf maple (*Acer macrophyllum*), black cottonwood (*Populus balsamifera* ssp. *trichocarpa*), and willows (*Salix* spp.).

Henry Cowell State Park, Santa Cruz County. Just south of Felton on Hwy 9 in the Santa Cruz Mountains, five miles north of the town of Santa Cruz. The main entry is near the San Lorenzo River, which is lined with superb specimens of western sycamore (*Platanus racemosa*), bigleaf maple, box elder (*Acer negundo* var. *californicum*), black cottonwood, and red alder with an understory of willows. Vines are mainly California blackberry (*Rubus ursinus*).

Mission Canyon Santa Barbara, Santa Barbara County. From Foothill Road in Santa Barbara, turn north onto Mission Canyon Road and follow it to the Santa Barbara Botanic Garden—which, besides being one of California's premier native botanical gardens, includes upper Mission Canyon, at the base of the Santa Ynez Mountains. Trees include California bay (*Umbellularia californica*), coast live oak (*Quercus agrifolia*), western sycamore, white alder (*Alnus rhombifolia*), and willows. The understory has California blackberry, foothill meadow-rue (*Thalictrum fendleri* var. *polycarpum*), canyon sunflower (*Venegasia carpesioides*), Humboldt lily (*Lilium humboldtii* var. *ocellatum*), canyon gooseberry (*Ribes menziesii*), melic grass (*Melica imperfecta*), common wood fern (*Dryopteris arguta*), and bracken fern (*Pteridium aquilinum*).

Inland Watercourses

Marsh Creek and Morgan Territory Regional Park, Contra Costa County. East of Clayton in Contra Costa County, take Clayton Road to Marsh Creek Road and follow it to the turnoff to Morgan Territory Road. Marsh Creek and Morgan Territory Regional Park have watercourses lined with bigleaf maple, white alder, western sycamore, Fremont cottonwood (*Populus fremontii*), and willows. Understory plants include foothill meadow-rue, California rose (*Rosa californica*), bracken fern, snowberry (*Symphoricarpos albus* var. *laevigatus*), California grape (*Vitis californica*), and vine honeysuckle (*Lonicera hispidula* var. *vacillans*).

Sespe Creek, Hwy 33, Ventura County. This area lies on the edge of the Sespe Wilderness Area and is accessed from Hwy 33 several miles north of Ojai. Trees include Fremont cottonwood, white alder, flowering ash (*Fraxinus dipetala*), and willows. Wildflowers include California rose, annual paintbrush (*Castilleja minor*), wild licorice (*Glycyrrhiza lepidota*), Indian hemp (*Apocynum cannabinum*), and false indigo (*Amorpha* sp.).

Sierra Watercourses

Deer Creek, east of Chico, Butte County. From Chico in the northern Sierra foothills, take Hwy 32 east toward Mt. Lassen. After passing through foothills and over a rise, the highway follows the Deer Creek drainage for several miles. Riparian trees include flowering dogwood (*Cornus nuttallii*), bigleaf maple, white alder, Fremont cottonwood, and willows, while the surrounding forest has white fir (*Abies concolor*), ponderosa pine (*Pinus ponderosa*), sugar pine (*P. lambertiana*), incense-cedar (*Calocedrus decurrens*), and California black oak (*Quercus kelloggii*). Associated understory plants include

umbrella plant (*Darmera peltata*), California rose, Sierra gooseberry (*Ribes roezlii*), native blackberry, blackcap raspberry (*Rubus leucodermis*), California grape, torrent sedge (*Carex nudata*), and brook orchid (*Epipactis gigantea*).

Rock Creek, Mono County. From Tom's Place near Sherwin Summit on Hwy 395 north of Bishop, turn west onto Rock Creek Road and continue to the end. The road closely follows Rock Creek. This is a wonderful example of high-elevation riparian woodland, including quaking aspen (*Populus tremuloides*), Fremont cottonwood, mountain alder (*Alnus incana* var. *tenuifolia*), and willows. Companion plants in-

clude monkshood (*Aconitum columbianum*), water hemlock (*Cicuta douglasii*), and swamp onion (*Allium validum*).

Kern River Canyon, east of Bakersfield, Kern County. Take Hwy 178 east from Bakersfield toward Lake Isabella and Walker Pass. The highway passes through the gorge of the lower Kern River for several miles. A stretch of old highway gives a more intimate view of the vegetation. Trees along the river include western sycamore, valley oak (*Quercus lobata*), white alder, Fremont cottonwood, and willows. Surrounding vegetation features chaparral, oak woodland, and scattered grassland.

Riparian woodland with willows (*Salix* ssp.) and western sycamore (*Platanus racemosa*) along Kern River, Kern Co.

CHAPTER 12

WETLANDS: THE BEAUTY OF WATER IN THE GARDEN

In a state noted for its parched, dry summer landscapes, water has surprising presence. Because we overuse water for development, industry, and agriculture, however, many of our natural water resources are nearly exhausted or greatly diminished. California once had vast wetlands—expansive sloughs and freshwater marshes; oxbow lakes and meanders in the Sacramento Delta; salt marshes lining bays and tidal flats; vernal pools set among foothill grasslands; pristine mountain and alpine lakes; and bogs, swamps, fens, and ponds.

Marshes—shallow freshwater wetlands—ringed by cattails (*Typha* spp.) and giant tules (*Scirpus* spp.), play host to immense migrations of water fowl. Swamps, bogs, and fens—wetlands of wooded places—harbor an assortment of unusual plants, including insectivorous plants that earn their living by trapping insects for food. Sparkling lakes fill cirques hollowed out by glaciers; these later silt up to become ponds and, finally, meadows. Vernal pools (also known as hogwallows) are depressions in foothill clays that fill with winter rains, then, in spring, erupt in rings of colorful wildflowers.

Each wetland has specialized plants, noted for their ability to thrive in soggy, oxygen-starved soils. They do so through several ingenious adaptations, such as hollow stems and leaf stalks, internal air chambers that permit them to float; and feathery underwater leaves, providing an expanded surface to absorb light, oxygen, and carbon dioxide. Such plants are beautifully adapted to water features in gardens. And happily, remnant wetlands remain in the wild to remind us of past glories as well.

LEFT: Pond ecosystem near Lassen National Park with watercress (*Rorippa* sp.) and lady ferns (*Athyrium filix-femina*). •
ABOVE: Bog with insectivorous cobra plants (*Darlingtonia californica*).

Creating a Garden

What better way to commemorate our remnant wetlands than to create a water feature in the garden? Even the smallest garden can include at least the suggestion of a pond or marsh—a wine barrel or urn filled with water, a simple fountain, or a bird bath. A variety of liners are now available for any size and depth of pond. Ponds enliven a garden and are an especially striking contrast to dryland plantings. Despite water loss from evaporation, ponds seldom use large amounts of water. Rather, they hint at a surplus of that most precious commodity.

Wetland plants run the gamut. Some ring the banks of marshes and ponds. Others live in shallow to deeper water, anchored to the mud by probing roots or sturdy rhizomes, sometimes splaying their leaves across the water's surface. A few float freely

on the surface and have no roots at all. Still others live completely submerged, only briefly revealing themselves when their flowers rise above the water.

Water plants require a couple of caveats. Most reproduce vigorously by creeping stems, quickly becoming invasive. Confining such plants to submerged boxes helps to slow their spread, and periodic thinning is necessary to avoid a formidable tangle. The free-floaters should not be considered at all; they multiply so rapidly that they will soon cover the pond's

TOP: Close up of garden water feature with great rush (*Juncus acutus*) and Catalina perfume (*Ribes viburnifolium*) in the foreground. • BOTTOM: Entry water garden. Photograph by G. Rubin.

surface, preventing oxygen exchange to underwater creatures such as fish, dragon fly larvae, and tadpoles.

If you have room for a pond several feet in diameter, you can create the different levels so common in natural ponds and marshes. Start by selecting plants for the perimeter, where soils remain moist but plants are not in standing water. Many herbaceous and woody perennials are suitable for this zone; most are winter dormant. On higher ground behind this zone, install a backdrop of complementary shrubs. Selection of material for this woody framework can be taken from the riparian palette discussed on page 279.

In shallow water at depths of eight to twelve inches, use the larger, grass-like plants such as tules (*Scirpus acutus*) and cattails (*Typha* spp.). Because of their height, you'll want to confine

these plants to one side of the pond to prevent blocking the view into the pond's center. At depths of two to three feet, grow the floating, anchored aquatics, including two water-lily relatives: yellow pond-lily (*Nuphar lutea* ssp. *polysepala*) and water-shield (*Brasenia schreberi*). The final touch is to install submerged aquatics such as water buttercup (*Ranunculus aquatilis*) and elodea (*Elodea canadensis*) in the lily zone. These good oxygenators help keep the water fresh for the animals that inhabit the pond.

Water gardens are seasonal, and most of the drama occurs from late spring to early fall, with a definite winter rest. Cut back the larger plants in winter. Maintain interest then by using evergreen sedges and rushes. For an evergreen backdrop, use

Natural pond with California rushes (*Juncus* sp.). Photograph by R. Driemeyer.

shrubs such as wax myrtle (*Myrica californica*), or feature deciduous shrubs with colorful twigs such as red-twig dogwood (*Cornus sericea*) and various willows (*Salix* spp.).

Although water gardens have many interesting foliage plants, colorful flowers can also be a part of the design. Besides the dramatic, waxy yellow blossoms of the pond-lily, there is a host of peripheral perennials with attractive flowers, including lemon lily (*Lilium parryi*), bog lupine (*Lupinus polyphyllus*), delta verbena (*Verbena hastata*), and delta marsh-mallow (*Hibiscus lasiocarpus*). Most bloom in the summer, when color is waning in other parts of the native garden.

Design Notes

This property, the site of a proposed remodel—a contemporary style, with a dramatic curved entry reminiscent of a ship's sail—is approached by a winding mountain road. As the direction of the slope changes, the plant communities create a mosaic, dominated by chaparral on south- and west-facing slopes and mixed-evergreen forest and oaks on north- and east-facing slopes. The garden itself was sited on a fairly steep hill with a northwest exposure. Unfortunately, the final phases of the project could not be completed; nevertheless, the design provides a good example of a wetland garden.

The property is less than twenty-five miles from the ocean. Although it does not lie in the coastal fog belt, it receives plentiful rainfall. The trees on-site—valley oak, coast life oak, bay, madrone, big-leaf maple, and redwood—are large, old, and healthy. The presence of riparian species and the size of the valley oak indicate a high water table as well.

Not long before my first visit, a drain line had been installed near the bottom of the slope, and the contractor had left a pile of soil and had excavated a deep ravine that was collecting runoff. I was able to observe the water action during a heavy storm. As I watched the water sheeting off the hill and heading toward the ravine, it occurred to me that I could create a series of rivulets to channel the water in the direction of the ravine, which would then serve as a retention basin. A couple of low fieldstone retaining walls with deep concrete foundations would hold the bank in place between the road and the basin. The rivulets, constructed of river cobbles, were graded to enhance the flow of water at various points along the new, saw-cut driveway. A seasonally active stream, with boulders and river-washed rocks and a waterfall, completed the scene.

In the ravine, water would have recirculated in an enclosed pond with a liner and pump. Directly adjacent to the pond lies a seasonal wetland, with water gradually percolating into the soil or flowing down the hill. Drains were installed to handle runoff in very wet years.

Because the owners desired a privacy screen and plants that provide fall color, the

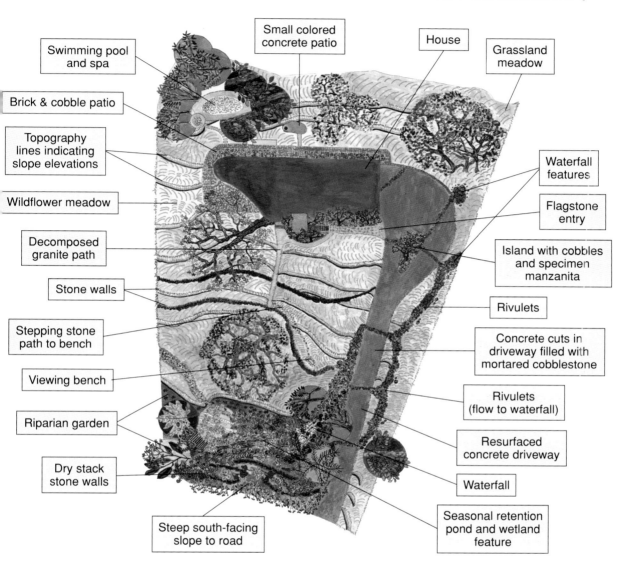

Swimming pool and spa

Brick & cobble patio

Topography lines indicating slope elevations

Wildflower meadow

Decomposed granite path

Stone walls

Stepping stone path to bench

Viewing bench

Riparian garden

Dry stack stone walls

Small colored concrete patio

Steep south-facing slope to road

House

Grassland meadow

Waterfall features

Flagstone entry

Island with cobbles and specimen manzanita

Rivulets

Concrete cuts in driveway filled with mortared cobblestone

Rivulets (flow to waterfall)

Resurfaced concrete driveway

Waterfall

Seasonal retention pond and wetland feature

wetland-riparian garden design features mountain species such as aspen trees, western azaleas, vine maples, and flowering currants, as well as another redwood. (Although these higher-elevation species do produce fall color, it is not as dramatic at sea level as in the mountains.) Above the wetland area, we intended to hydroseed a native grassland meadow as understory to the existing oaks. At the edge of the road it is hot, sunny, and dry, ideal for low-growing chaparral species. Finally, manzanita would be massed at the entry to the house, and one larger specimen planted in a bed of cobbles in the driveway island.

Rendering of featured garden.

Retaining rainwater on site is a strong component of our garden-making philosophy. Although it is a creative challenge to accomplish this on steep slopes, it is worth the effort to preserve the watershed.

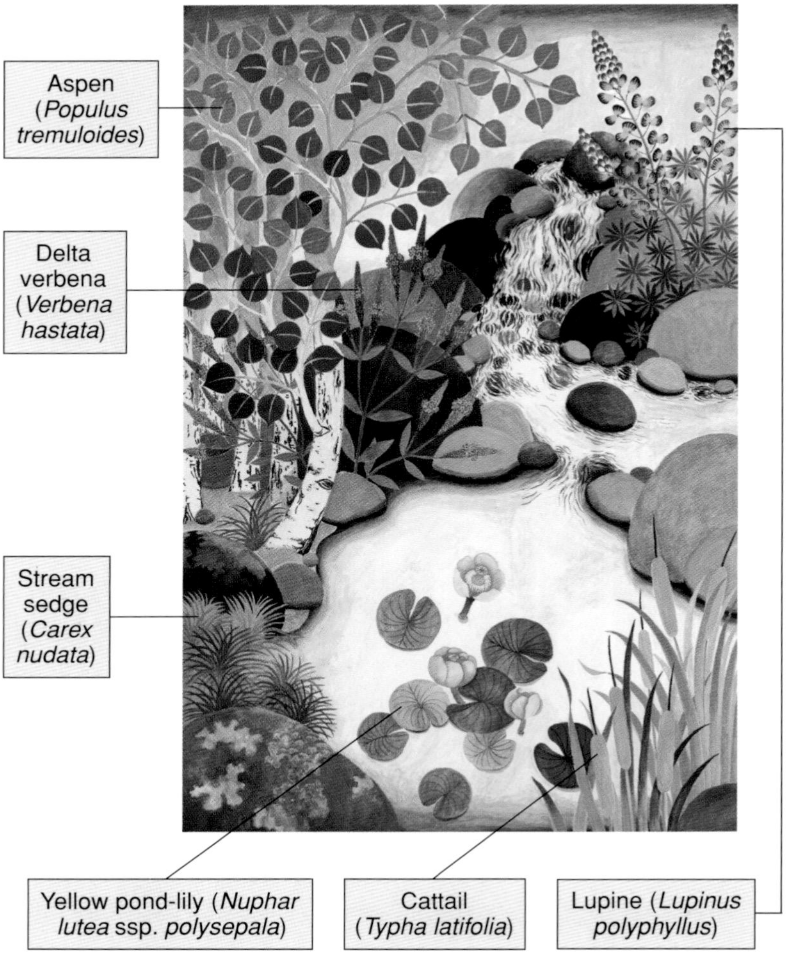

Scope of Work

* Finish grade to channel runoff to water features.
* Create a drainage system that allows slow, even distribution of runoff as well as the ability to percolate. Install street drains too, for when soils become saturated.
* Build retaining walls, with drainage behind.

Aspen (*Populus tremuloides*)

Delta verbena (*Verbena hastata*)

Stream sedge (*Carex nudata*)

Yellow pond-lily (*Nuphar lutea* ssp. *polysepala*)

Cattail (*Typha latifolia*)

Lupine (*Lupinus polyphyllus*)

* Create enclosed ponds with flexible vinyl lining and recirculating pump.
* Build seasonal creek water feature with rivulets, waterfall, and boulders.
* Create other decorative fieldstone walls.
* Build decomposed granite pathways and install a bench.
* Install drip and overhead spray irrigation.
* Install plants.
* Install invasive water plants in the pond feature in plastic containers. Secure them in place.
* Mulch all planted areas.
* Hydroseed meadow areas.
* Begin aggressive weeding program when meadow seeds germinate.

Featured garden artwork. Illustration by A. Yankellow.

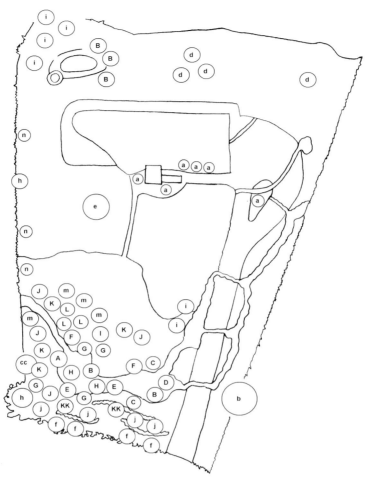

Plant List

SYMBOL	BOTANICAL NAME	COMMON NAME

Wetlands

SYMBOL	BOTANICAL NAME	COMMON NAME
A	*Populus tremuloides*	quaking aspen
B	*Carex spissa*	blue-leafed sedge
C	*Lupinus polyphyllus*	bog lupine
D	*Nuphar lutea* ssp. *polysepala*	yellow pond-lily
E	*Typha angustifolia*	narrow leaf cattail
F	*Verbena hastata*	delta verbena

Riparian

SYMBOL	BOTANICAL NAME	COMMON NAME
G	*Carex nudata*	torrent sedge

Featured garden key.

H	*Juncus patens*	California rush
I	*Philadelphus lewisii*	western mock-orange
J	*Physocarpus capitatus*	ninebark
K	*Polystichum munitum*	western sword fern
L	*Woodwardia fimbriata*	giant chain fern

Mixed-Evergreen Forest/Oak Woodland

a	*Arctostaphylos manzanita*	common manzanita
b	*Arbutus menziesii*	madrone
cc	*Acer macrophyllum*	big-leaf maple
d	*Quercus agrifolia*	coast live oak
e	*Q. lobata*	valley oak
f	*Salvia sonomensis*	Sonoma sage
g	*Sequoia sempervirens*	coast redwood
h	*Heteromeles arbutifolia*	toyon
i	*Umbellularia californica*	California bay

Chaparral

j	*Ceanothus maritimus*	maritime ceanothus
KK	*Fremontodendron californicum* ssp. *decumbens*	prostrate flannelbush
LL	*Myrica californica*	California wax myrtle

Redwood Forest

m	*Berberis aquifolium* var. *repens*	creeping barberry
n	*Myrica californica*	California wax myrtle
o	*Ribes sanguineum* var. *glutinosum*	pink-flowering currant

Wildflower Meadow (from seed)

	Festuca idahoensis	Idaho fescue
	Melica californica	California melic
	Nassella pulchra	purple needlegrass
	Achillea millefolium	yarrow
	Clarkia amoena	farewell-to-spring
	Eschscholzia californica	California poppy
	Iris douglasiana	Douglas iris
	Triteleia laxa 'Queen Fabiola' (bulb)	Ithuriel's spear

Plants to Use

Note: Although the selections below are not discussed in other chapters, several riparian species (particularly shade plants) and also some desert oasis species are appropriate in water gardens. Note that many of these plants are available only from specialty nurseries or at special plant sales.

Shoreline Plants

Perennials

Indian hemp (*Apocynum cannabinum*).

Invasive, winter-dormant, herbaceous perennial to six feet tall. Dense colonial stems from creeping rhizomes, with broadly elliptical leaves in pairs and cymes of small white flowers in summer. Leaves turn bright yellow in fall. Flowers attract butterflies. Propagate from root divisions. Cut back old stems to the ground in winter.

Delta verbena (*Verbena hastata*).
Winter-dormant, herbaceous perennial to five feet tall. Narrow clumps of stems with pairs of toothed, ovate leaves and narrow spikes of purple, bee-pollinated flowers through summer. Good substitute for the popular, nonnative *V. bonariensis* common in perennial borders. Propagate from seed or root divisions. Cut old stems to the ground in winter. Garden Design Note: Can be invasive. Seeds germinate easily.

California marsh-mallow (*Hibiscus lasiocarpus*).
Winter-dormant, woody perennial to four feet high. Broadly ovate, toothed leaves and large, mallowlike, white flowers with a dark red center

TOP RIGHT: Masses of flowers on Indian hemp (*Apocynum cannabinum*). • BOTTOM LEFT: Flowering spikes of delta verbena (*Verbena hastata*).

in late summer. Propagate from seed, cuttings, or divisions. Prune back old stems in winter.

Bog lupine (*Lupinus polyphyllus*).
Winter-dormant, herbaceous perennial with large, palmately compound leaves. Two to three feet high when not in bloom; flower spikes add another foot or two. Dense racemes of purple to deep blue, sweet pea-like flowers in summer are followed by silky-haired seed pods. Propagate from fresh seed. Cut back stems in winter.
Garden Design Note: Parent of the Russell hybrids. Very desirable in a wetland garden.

Coast woodmint (*Stachys chamissonis*).

TOP: Bog lupine (*Lupinus polyphyllus*). • BOTTOM: Coast woodmint (*Stachys chamissonis*).

Invasive, winter-dormant, herbaceous perennial to six feet tall. Pairs of quilted, strongly scented, elliptical leaves; narrow spikes of rose-purple flowers attractive to many different pollinators in summer. Propagate from divisions of rhizomes. Cut back stems in winter. Be sure to allot plenty of space for roots to spread.

Garden Design Note: Hummingbirds like this plant.

Cardinal flower or scarlet lobelia (*Lobelia cardinalis*).
Clump-forming, winter-dormant perennial to three feet tall in flower. Lance-elliptical leaves (some cultivars have dark red leaves) and spikes of showy, two-lipped, scarlet flowers in summer attractive to hummingbirds. Propagate from division of clumps or seed.

Garden Design Note: Full sun and wet conditions.

Lemon lily (*Lilium parryi*).
Winter-dormant bulb to four feet tall. Whorls of narrow leaves and racemes of large, highly fragrant, waxy yellow, trumpetlike lily flowers in early summer. Propagate from seed (three years to bloom) or bulb scales.

Western skunk cabbage (*Lysichiton americanum*).
Robust, tropical-looking, winter-dormant perennial to three feet high. Bold, oval, bright green leaves; robust spikes of tiny whitish flowers surrounded by a showy, yellow, petal-like bract in early to midspring. Easily propagated from seed, or divide clumps of roots. Cut off old flowering stalks unless volunteer seedlings are wanted.

Garden Design Note: Likes wet, soggy bogs. Available from Oregon nurseries.

Grasslike Plants

Southern rush (*Juncus acutus*).
Good year-round foliage plant with clumps of stiffly vertical, dark green, sharply pointed stems. Grows to six feet tall. Insignificant brownish flowers in summer. Propagate from seed or divisions. Divide clumps every few years to remove old growth and prevent entangled masses of roots. Be sure to use care around the spiny stems.

TOP: Scarlet lobelia (*Lobelia cardinalis*). • BOTTOM: Skunk cabbage (*Lysichiton americanum*) in a redwood wetland.

Garden Design Note: Excellent container plant. Keep away from eyes by placing it high above an entry. Good vertical element, and makes a fine focal point in the modern garden.

Fiber optic rush (*Scirpus cernuus*).
Miniature, clump-forming perennial with masses of bright green, rushlike stems that resemble green fiber optic cable. Plants grow no more than six inches high. Insignificant spikes of wind-pollinated flowers. Propagate from divisions.
Garden Design Note: Tuck it between rocks near a waterfall.

Blue-leaved or San Diego sedge (*Carex spissa*).
Semidormant, clumped, grasslike plant to six feet tall with stiff, rough, bluish green leaves. Dramatic spikes of brownish to pale yellow flowers appear in summer. Propagate from divisions or seed. Divide clumps every few years and remove dead growth in winter.
Garden Design Note: Can take less water. When in flower, it is stunning. Grows fast and large.

Ground Covers

Swale checkermallow (*Sidalcea calycosa* ssp. *rhizomata*).
Rhizomatous, winter-dormant, creeping, herbaceous perennial. Mostly basal, rounded, toothed to palmately lobed leaves until bloom time; racemes of purple, mallowlike flowers in summer. Propagate from divisions of rhizomes. Lift and divide clumps every few years to prevent tangling of rootstocks and roots. The better known *S. calycosa* spp. *calycosa* is a beautiful, late spring-flowering annual from vernal pools.
Garden Design Note: Difficult to find.

Creeping waterpea (*Hoita orbicularis*).
Widely wandering, invasive, winter-dormant, herbaceous perennial. Large, beanlike, trifoli-

TOP LEFT: Fiber optic rush (*Scirpus cernuus*) in a container fountain accented with cast bronze acorns by artist Ben Hunt. Photograph by S. Ingram. • BOTTOM RIGHT: Ground cover of creeping waterpea (*Hoita orbicularis*).

ate leaves on stalks to a foot high, with two-foot-tall racemes of pink-purple, sweet pea-like flowers in early summer. Be sure to provide ample space for plants to spread; stolons are tough, wiry, and difficult to eradicate.

Yerba mansa (*Anemopsis californica*).
Winter-dormant, herbaceous, perennial ground cover with clumps of oblong leaves and red, strawberrylike runners. Fragrant, conelike, white flowers in summer. Propagate from plantlets on runners. May become invasive.
Garden Design Note: This plant was used by indigenous people for medicinal purposes.

Shallow-Water Plants

Grasslike plants

Cattails (*Typha* spp.).
Vigorous, rhizomatous, winter-dormant, grasslike perennials with narrow, bluish green leaves to four feet high. Stout stems bear double, sausage-shaped flower spikes in early

TOP: Flower of yerba mansa (*Anemopsis californica*). • BOTTOM: Plants with fruiting spikes of common cattail (*Typha latifolia*).

summer. The lower spike ripens into a brown cylinder in fall, releasing thousands of white-haired seeds. Propagate from divisions of rhizomes. Lift and divide clumps every couple of years in winter.

Garden Design Note: Excellent choice for a containerized water feature. Dies back to the ground.

California tule (*Scirpus acutus*).
Winter-dormant, grasslike perennial with very tall, strongly vertical, dark green stems to twelve feet high (leaves are reduced to scales). Cymes of tiny, greenish flowers near the tops of stems in summer. Propagate from divisions of rhizomes. Divide clumps every few years.

Garden Design Note: Use like cattails.

Ground Covers

Silverweed (*Potentilla anserina* ssp. *pacifica*).
Creeping, invasive, winter-dormant, herbaceous, perennial ground cover that grows in shallow water or moist soil. Spreads by red, strawberry-like runners. Rosettes of pinnately compound, silver-backed leaves. Single, yellow, roselike flowers from spring through summer. Propagate from plantlets on runners. Allow plenty of room for plants to spread; roots are sturdy and difficult to remove.

Garden Design Note: Plant in a wet meadow or around a pond. Sun or shade.

Buckbean (*Menyanthes trifoliata*).
Creeping, winter-dormant, herbaceous perennial with large, trifoliate, beanlike leaves.

Silverweed (*Potentilla anserina* ssp. *pacifica*).

Short, narrow racemes of fringed, white to pale purple flowers in early summer. Propagate from rooted sections of rhizomes.

Marsh cinquefoil (*Potentilla palustris*).
Creeping, winter-dormant, herbaceous perennial with pinnately compound leaves and short racemes of dark red, roselike flowers in summer. Propagate from sections of rhizomes.

Perennials

Arrowhead (*Sagittaria latifolia*).
Tuberous, winter-dormant, herbaceous perennial rooted in mud. Large, arrowhead-shaped leaves grow partly above water, with racemes of white flowers emerging in summer. Propagate from small tubers.

Floating Plants

Yellow pond-lily (*Nuphar lutea* ssp. *polysepala*).
Winter-dormant, herbaceous perennial with stout rhizomes anchored in mud. Sturdy, round leaves rise just

above water or float on the surface. Large, emergent, waxy, yellow, waterlily-like flowers appear in summer. Propagate from divisions of rhizomes. Lift and divide rhizomes to avoid tangling every couple of years in winter.

Water shield (*Brasenia schreberi*).
Winter-dormant, herbaceous perennial from fleshy roots anchored in mud. Slimy, shield-shaped leaves float on water surface, with emergent, dark red-purple flowers appearing in summer. Propagate from offsets.

Pondweed (*Potamogeton* spp.).
Winter-dormant, herbaceous perennials from slender rhizomes anchored in mud. Oblong, reddish-tinted floating leaves and narrow, grasslike submerged leaves. Insignificant spikes of greenish flower racemes emerge above water in summer. Propagate from divisions of rhizomes. Thin plants every couple of years.

LEFT: Yellow pond-lily (*Nuphar lutea* ssp. *polysepala*). • RIGHT: Leaves of arrowhead (*Sagittaria latifolia*).

Submerged Aquatics

Water buttercup (*Ranunculus aquatilis*).
Winter-dormant, herbaceous perennial with finely divided, feathery leaves under water. Occasional floating, palmately lobed leaves. Emergent snowy white flowers in late spring to early summer. Propagate from rooted sections of stems. Thin plants annually in winter.

Elodea (*Elodea canadensis*).
Winter-dormant, herbaceous perennial with filmy, lance-shaped leaves under water.

Small, white, emergent flowers in summer. Excellent oxygenator. Propagate from rooted sections of stems. Thin plants annually in winter.

Bladderwort
(*Utricularia vulgaris*).
Winter-dormant, herbaceous perennial with delicate, feathery, submerged leaves and tiny, dark, underwater bladders that trap insect larvae. Emergent, yellow, spurred, snapdragon-like flowers appear in summer. Propagate from rooted sections of stems.

Additional shoreline plants to try:

Spike-rushes (*Eleocharis* spp.), blue lobelia (*Lobelia dunnii*), seep monkeyflower

Siesta Lake, a shallow pond with yellow pond-lily (*Nuphar lutea* ssp. *polysepala*) in Yosemite National Park.

(*Mimulus guttatus*), scarlet monkeyflower (*M. cardinalis*), fragrant pluchea (*Pluchea odorata*), common rush (*Juncus patens*), leopard lily (*Lilium pardalinum*), California yampah (*Perideridia californica*), leather-root (*Hoita macrostachya*), western goldenrod (*Euthamia occidentalis*), lady fern (*Athyrium filix-femina*), giant chain fern (*Woodwardia fimbriata*).

Additional shallow-water plants to try:

Water-primrose (*Ludwigia* spp.), water-plantain (*Alisma plantago-aquatica*), mare's tail (*Myriophyllum* spp.), clover fern (*Marsilea vestita*), bur-reed (*Sparganium* spp.), watercress (*Rorippa nasturtium-aquaticum*).

Additional floating plants to try:

Water knotweed (*Polygonum coccineum*).

Additional submerged plants to try:

Water hornwort (*Ceratophyllum demersum*).

TOP: Modern fountain with stainless steel façade. Photograph by R. Driemeyer. • BOTTOM: Pond with water birch (*Betula occidentalis*) in the background and yellow pond-lily (*Nuphar lutea* ssp. *polysepala*) and sedges (*Carex* spp.) in the water.

Places to Visit

Wetlands vary greatly and are widely scattered through the state. Suggested sites to visit are divided into four categories, though others exist as well: ponds, bogs, marshes, and vernal pools.

Ponds

False Klamath Cove, Del Norte County. A few miles north of Requa on Hwy 101 by the coast. This cove was mistaken for the outlet to the Klamath River, which lies a few miles to the south. It features a pond-bog ecosystem surrounded by north coastal coniferous forest and dunes. Plants along the perimeter of the cove include steeple bush spiraea (*Spiraea douglasii*), willows (*Salix* spp.), yellow pond-lily (*Nuphar lutea* ssp. *polysepala*), water parsley (*Oenanthe sarmentosa*), pondweed (*Potamogeton* sp.), common daisy (*Erigeron philadelphicus*), and marsh cinquefoil (*Potentilla palustris*).

Nipomo Dunes, San Luis Obispo County. West of Nipomo and south of Arroyo Grande off Hwy 1 in southern San Luis Obispo County. The Nature Conservancy administers Oso Flaco Preserve, which includes giant coastal sand dunes, brackish and freshwater lakes, and coastal scrub. The lakes are home to water parsley, tules (*Scirpus* spp.), sedges (*Carex* spp.), rushes (*Juncus* spp.), water knotweed (*Polygonum* sp.), silverweed (*Potentilla anserina* var. *pacifica*), and yellow pond-lily.

Grass Lake, Alpine County. This large meadow-pond-bog ecosystem lies at the top of Luther Pass on Hwy 89 just south of Lake Tahoe. The open wetlands are surrounded by dense lodgepole pine forest. Grass Lake is home to sedges, cotton-grass (*Eriophorum gracile*), buckbean (*Menyanthes trifoliata*), marsh cinquefoil, yellow pond-lily, western bilberry (*Vaccinium uli-*

ginosum ssp. *occidentale*), sundew (*Drosera rotundifolia*), primrose monkeyflower (*Mimulus primuloides*), alpine shooting stars (*Dodecatheon alpinum*), common bladderwort (*Utricularia vulgaris*), and pondweed.

Bogs

Old Gasquet Toll Road, Del Norte County. From the town of Gasquet on Hwy 199 northeast of Crescent City, turn north onto Gasquet Toll Road. Drive several miles. Sphagnum bogs lie along the road among mixed coniferous forest and northern serpentine chaparral. Bog plants include narthecium (*Narthecium californicum*), bog asphodel (*Tofieldia glutinosa*), Vollmer's lily (*Lilium pardalinum* ssp. *vollmeri*), California pitcher plant (*Darlingtonia californica*), sundew, California ladyslipper (*Cypripedium californicum*), Labrador tea (*Ledum glandulosum*), western azalea (*Rhododendron occidentale*), meadow paintbrush (*Castilleja miniata*), and California coneflower (*Rudbeckia californica* var. *glauca*).

Butterfly Valley near Quincy, Plumas County. From Hwy 70 to Quincy, turn south around five miles west of Quincy near the town of Keddie. You need a local map to interpret the back roads here—Butterfly Valley is not on a clearly marked road. The valley is a series of wet meadows and sphagnum bogs surrounded by mid-elevation mixed conifer forest. The open areas are home to camas (*Camassia quamash*), leopard lily (*Lilium pardalinum*), narthecium, bog asphodel, California pitcher plant, yampah (*Perideridia* sp.), white meadow violet (*Viola macloskeyi*), sundew, Labrador tea, western azalea, steeple bush spiraea, cattail (*Typha latifolia*), arrowhead (*Sagittaria latifolia*), common bladderwort, sedges, and water-shield (*Brasenia schreberi*).

Marshes

Drake's Estero, Pt. Reyes National Seashore, Marin County. Drake's Bay lies on a spur road on the south side of Sir Francis Drake Boulevard, which departs from Hwy 1 near the town of Pt. Reyes Station. The parking lot sits between coastal bluffs, a small beach area, and, at the far end, a freshwater marsh. The marsh is ringed with willows and features the classic zones from marsh edge to relatively deep water. Plants include cattail, tule, silverweed, water knotweed (*Polygonum* sp.), duckweed (*Lemna* sp.), duck fern (*Azolla filiculoides*), rushes, and sedges.

Pescadero Marsh, San Mateo County. This large wetland area lies behind dramatic dunes on the San Mateo coast south of Half Moon Bay near the tiny town of Pescadero. From Hwy 1 you can access a trail system into these wetlands, which feature salt, brackish, and freshwater marshes. (The inland portion is freshwater.) The marshes are home to many different sedges, tules, cattails, rushes, silverweed, water knotweed, and others.

Boardwalk through freshwater marsh, Oso Flaco Dunes, San Luis Obispo Co.

Vernal pools

Note: The once extensive marshes in the Central Valley have mostly been drained.

Table Mountain and Larkin Road, Butte County. These areas lie just north and west of Oroville in the northern Sierra foothills. Table Mountain is accessed from Cherokee Road; the top lava barrens have depressions that become vernal pools in winter and spring. Larkin Road goes south from Hwy 160 west of Oroville; near Clay Pit Mine is a meandering, unprotected, clay-based vernal pool. These vernal pools are home

to two or three different downingias, annual checkerbloom (*Sidalcea calycosa*), annual golden monkeyflower (*Mimulus guttatus*), vernal pool navarretia (*Navarretia leucocephala*), snowy meadowfoam (*Limnanthes douglasii* var. *niveus*), fragrant clover (*Trifolium variegatum*), yellow gentian (*Cicendia quadrangularis*), water buttercup (*Ranunculus aquatilis*), and button parsley (*Eryngium* sp.).

Jepson Prairie, Solano County. Jepson Prairie lies southeast of Fairfield and Vacaville just north of Hwy 12 off Hwy 113. The area comprises bunchgrasslands and a series of small to large vernal pools. Special plants include white fritillary (*Fritillaria liliacea*), blue dicks (*Dichelostemma capitatum*), owl's clover (*Castilleja exserta*), cream sacs (*Triphysaria eriantha*), glueseed (*Blennosperma nanum*), goldfields (*Lasthenia* spp.), a couple of rare vernal pool grasses (*Neostapfia colusana* and *Orcuttia* sp.), and three species of *Downingia*.

Redhills area near Chinese Camp, Tuolumne County. Take Redhills Road southwest from Chinese Camp. The red hills are formed of red serpentine rocks, with small vernal pools and streams among chaparral, foothill woodland, and grassland. Chinese Camp is a tiny town south of Sonora on Hwy 49/120 in the cental Sierra foothills. The vernal pools here feature popcorn flower (*Plagiobothrys* sp.), button parsley, pallid brodiaea (*Brodiaea pallida*), annual golden monkeyflower, ladies' tresses (*Spiranthes* sp.), goldfields, meadowfoam, and more.

Kearny and other mesas on the north edge of San Diego, San Diego County. Although rapid suburban development has hit this area, remnant habitats still support vernal pools. Among the typical species are San Diego Mesa mint (*Pogogyne abramsii*), Loma Alta mint (*P. nudiuscula*), popcorn flower (*Plagiobothrys acanthocarpus*), downingia (*Downingia cuspidata*), San Diego coyote thistle (*Eryngium aristulatum* ssp. *palmeri*), Orcutt's brodiaea (*Brodiaea orcuttii*), and woolly heads (*Psilocarphus tenellus*).

Vernal pool with *Sidalcea calycosa* (pink) and the nonnative leon's bit (*Leontodon taraxacoides*, yellow) in the foothills of the northern Sierra near Oroville.

APPENDIX 1

SOURCES OF NATIVES

One of the challenging and often daunting tasks of creating a native garden is finding appropriate plant material. Many species featured in this book can be found in the trade. Nonetheless, several of the plants named, especially those included in the lists of additional species to try, are hard to find.

Besides the commercial sources listed below, you can seek out special plant sales, which are often held annually. Prominent among these are the sales and "cutting days" sponsored by the many different chapters of the California Native Plant Society. Some chapters hold sales in the fall, the season for installing plants in the garden; others have spring sales, when the plants look their best.

Listed below are sources of plants and seeds. Compiled with the help of Friends of Regional Parks Botanic Garden in Tilden Regional Park, Berkeley, this list is subject to change over time. Another good source of information is the California Native Plant Link Exchange, www.cnplx.info.

Agua Fria Nursery
1409 Agua Fria St., Santa Fe NM 87501
(505) 983-4831
www.aguafrianursery.com
Mail order retail plants; wide selection, including uncommon penstemons; many California-collected natives.

Albright Seed Co.
189 Arthur Rd., Martinez CA 94553
(925) 372-8245
www.albrightseed.com
Bulk sales; grass, wildflower, shrub, and tree seed; 50 percent native.

Annie's Annuals
740 Market Ave., Richmond CA 94801
(510) 215-1671, retail; (510) 215-1326, wholesale
www.anniesannuals.com
Many native annual and perennial plants, some hard to find, in four-inch pots.

Appleton Forestry Nursery
1369 Tilton Rd., Sebastopol CA 95472
(707) 823-3776
Container trees and shrubs; contract growing. Call ahead.

Bayview Gardens
1201 Bay St., Santa Cruz CA 95060
(831) 423-3656
http://thegardensite.com/irises/bayviewgardens/
Mail order iris, featuring the Pacific Coast hybrids of Joe Ghio.

Berkeley Horticultural Nursery
1310 McGee Ave., Berkeley CA 94703
(510) 526-4704
www.berkeleyhort.com
Retail plants, with one section devoted to natives.

C. H. Baccus
900 Boynton Ave., San Jose CA 95117
(408) 244-2923
Mail order bulbs, especially calochortus, fritillaria.

Calaveras Nursery
1622 E. Hwy 12, Valley Springs CA 95252
(209) 772-1823
Retail trees, about 50 percent native.

California Flora Nursery
P.O. Box 3, Somers & D Sts., Fulton CA 95439
(707) 528-8813
www.calfloranursery.com
Wholesale and retail, native and Mediterranean plants; no shipping or mail order.

California Department of Forestry, I. Moran Reforestation Center
5800 Chiles Rd., Davis CA 95617
(530) 753-2441
Tree sales for erosion control, windbreaks, and reforestation; 50 items minimum order.

Central Coast Wilds
336 Golf Club Dr., Santa Cruz CA 95060 (nursery)
114 Liberty St., Santa Cruz CA 95060 (mailing address)
(831) 459-0655
www.centralcoastwilds.com
An ecologically sensitive native plant nursery; ecological restoration and botanical consulting.

Circuit Rider Productions Inc.
9619 Old Redwood Hwy, Windsor CA 95492
(707) 838-6641
www.crpinc.org
By appointment only; wholesale and retail plants; contract collecting and growing; revegetation and reforestation.

Clyde Robin Seed Co.
P.O. Box 2366, Castro Valley CA 94546
(510) 785-0425
http://clyderobin.com
Wholesale and mail order seed.

ConservaSeed
P.O. Box 455, Rio Vista CA 94571
(916) 775-1676
Wholesale grass, forb, and legume seed; contract collecting and growing; revegetation and restoration.

Cornflower Farms
P.O. Box 896, Elk Grove CA 95759
(916) 689-1015
www.cornflowerfarms.com
Wholesale only; container plants, all sizes; 80-plus percent native; revegetation and restoration.

Elkhorn Native Plant Nursery
P.O. Box 270, 1957B Hwy 1, Moss Landing CA 95039
(831) 763-1207
www.elkhornnursery.com
Wholesale and retail; seed, container and bareroot plants; contract collecting and growing; demo garden.

Far West Bulb Farm
14499 Lower Colfax Rd., Grass Valley CA 95945
(530) 272-4775
www.californianativebulbs.com
Mail order bulbs native to central and northern California; contract growing.

Forestfarm
990 Tetherow Rd., Williams OR 97544
(541) 846-7269
http://forestfarm.com
Retail mail order plants; extensive selection of western trees and shrubs.

Forest Seeds of California
1100 Indian Hill Rd., Placerville CA 95667
(530) 621-1551
Mail order tree and shrub seeds; contract collecting.

Freshwater Farms Inc.
5851 Myrtle Ave., Eureka CA 95503
(707) 444-8261 or (800) 200-8969
www.freshwaterfarms.com
Wholesale and retail seed, container and bareroot riparian plants; contract collecting and growing, revegetation and restoration.

Greenlee Nursery
301 E. Franklin Ave., Pomona CA 91766
(909) 629-9045
www.greenleenursery.com
Wholesale and retail ornamental grasses, some native; mail order.

Hedgerow Farms
21740 County Rd. 88, Winters CA 95692
(530) 662-6847
www.hedgerowfarms.com
Wholesale and retail seed; container grasses, sedges, rushes, and forbs; contract collecting & growing.

Industrial Forest Associates
4886 Cottage Grove Ave., McKinleyville CA 95519
(707) 839-3256
Retail; bulk sales of native conifers.

Lake County Natives
7480 Kelsey Creek Dr., Kelseyville CA 95451
(707) 279-2868
Wholesale and retail plants; by appointment.

Larner Seeds
P.O. Box 407, Bolinas CA 94924
(415) 868-9407
www.larnerseeds.com
Mail order seeds, retail plants at nursery Oct.–July; coastal ecotypes of wildflowers, perennials, shrubs, trees, and grasses; demo garden; contract growing.

Las Pilitas Nursery
3232 Las Pilitas Rd., Santa Margarita CA 93453
(805) 438-5992
www.laspilitas.com
Wholesale, retail by appointment; seed and container plants; contract collecting and growing.

Matilija Nursery
8225 Waters Rd., Moorpark CA 93021
(805) 523-8604
Natives only; open to the public.

Moon Mountain Wildflowers
P.O. Box 725, Carpinteria CA 93014
(805) 684-2565
Mail order seed.

Mostly Natives Nursery
P.O. Box 258, 27235 Hwy One, Tomales CA 94971
(707) 878-2009
www.mostlynatives.com
Wholesale and retail plants; coastal natives and drought-tolerant plants.

Native Here Nursery
101 Golf Course Dr., Tilden Regional Park, Berkeley CA 94708
(510) 549-0211
www.ebcnps.org/ NativeHereHome.htm
Volunteer-run by the California Native Plant Society; excellent local native plants; revegetation and restoration; contract collecting and growing. Open Friday 9–noon and Saturday 10–1 only; call ahead.

Native Revival Nursery
8022 Soquel Dr., Aptos CA 95003
(831) 684-1811
www.nativerevival.com
Wholesale and retail seed and plants; contract collecting and growing; revegetation and restoration.

Native Sons
379 West El Campo Rd., Arroyo Grande CA 93420
(805) 481-5996
www.nativeson.com
Wholesale; Mediterranean-climate plants and natives.

North Coast Native Nursery
P.O. Box 744, Petaluma CA 94953
(707) 769-1213
www.northcoastnativenursery.com
Native plants for woodland, coastal, and riparian habitats; wholesale and retail seed and plants; contract collecting and growing.

Northwest Native Seed
17595 Vierra Canyon Rd. #172, Prunedale CA 93907
Catalog issued late fall, ships seed Nov.–Apr.; extensive listing includes data on where collected; many hard-to-find plants. Ron Ratko, proprietor.

Pacific Coast Seed
833 Hawthorne Pl, Livermore CA 94551
(925) 373-4417
www.pcseed.com
Wholesale or through local nurseries; seed for wildflowers, shrubs, grasses, and trees.

Payless Nursery
2927 S. King Rd., San Jose CA 95122
(408) 274-7815
Permanent natives section with a selection that changes from month to month.

Plants of the Southwest
3095 Agua Fria Rd., Santa Fe NM 87507
(505) 438-8888 or (800) 788-7333
www.plantsofthesouthwest.com
Mail order plants and seeds, many native to California.

Rana Creek Habitat Restoration
35351 E. Carmel Valley Rd., Carmel Valley CA 93924
(831) 659-3820
www.ranacreeknursery.com
Wholesale only; revegetation; seed, container, and bareroot plants.

Rancho Santa Ana Botanic Garden
1500 N. College Ave., Claremont CA 91711
(909) 625-8767
www.rsabg.org
Natives from all over the state; design gardens as well as examples of California's major plant communities.

Regional Parks Botanic Garden
South Park Dr. & Wildcat Canyon Rd., Tilden Regional Park, Berkeley CA 94708
(510) 841-8732
www.nativeplants.org
Retail plants most Thursday mornings; seed in visitors center fall and winter; annual plant sale third Saturday in April.

Russell Graham, Purveyor of Plants
4030 Eagle Crest Rd. NW, Salem OR 97304
(503) 362-1135
Retail and mail order; 10 percent native, including hard-to-find plants such as Trillium.

Santa Barbara Botanic Garden
1212 Mission Canyon Rd., Santa Barbara CA 93105
(805) 682-4726
www.sbbg.org
Retail container plants, some hard-to-find natives.

Seedhunt
P.O. Box 96, Freedom CA 95019
www.seedhunt.com
Mail order annual and perennial seed; about one-third native, many hard to find.

Sierra Azul Nursery and Gardens
2660 E. Lake Ave. (Hwy 152), Watsonville CA 95076
(831) 753-0939
www.sierraazul.com
Mediterranean and water-conserving plants, some native.

Siskiyou Rare Plant Nursery
2115 Talent Ave., Talent OR 97540
(541) 535-7103
www.siskiyourareplantnursery.com
Retail and mail order; alpine and rock garden plants, including some hard-to-find natives.

Sonoma Horticutural Nursery
3970 Azalea Ave., Sebastopol CA 95472
(707) 823-6832
www.sonomahort.com
Woodland garden plants; known for rhododendrons and azaleas; some natives.

Southwestern Native Seeds
P.O. Box 50503, Tucson AZ 85703
Mail order seeds; some collected in California are hard to find.

Suncrest Nurseries Inc.
400 Casserly Rd., Watsonville CA 95076
(831) 728-2595
www.suncrestnurseries.com
Wholesale; see website for local retail outlets. Development of new and unusual ornamental plants for coastal California; many natives.

Telos Rare Bulbs
P.O. Box 4147, Arcata CA 95518
www.telosrarebulbs.com
Many natives including Calochortus, Brodiaea, Fritillaria, *and* Erythronium.

Theodore Payne Foundation
10459 Tuxford St., Sun Valley CA 91352
(818) 768-1802
www.theodorepayne.org
*Retail plants, mail order seed; large selection of Califor-
nia native plants and wildflowers.*

Tree of Life Nursery
P.O. Box 635, 33201 Ortega Hwy, San Juan
Capistrano CA 92693
(949) 728-0685
www.treeoflifenursery.com
*Wholesale and retail container plants; contract collect-
ing and growing.*

Yerba Buena Nursery
19500 Skyline Blvd., Woodside CA 94062
(650) 851-1668
www.yerbabuenanursery.com
Retail plants and some seed; large demo garden; except for ferns, all natives.

APPENDIX 2

BOOKS AND RESOURCES FOR LEARNING ABOUT NATIVES

One of the greatest challenges facing the gardener and designer is learning the names, habits, and habitats of natives. This book only hints at the richness of our vast flora, so there's great opportunity to
further your knowledge. Even with familiar plants, appropriate garden use hinges on knowing the plant
from seed to maturity in a variety of situations. The best way of increasing your knowledge is to visit native
habitats (see Places to Visit in each chapter), private and public gardens that feature natives, and websites,
and to read books that deal with natives. Here are a few of the best examples (note that open hours may
change, and many gardens are closed on holidays, so it's wise to check the website or call ahead):

BOTANIC GARDENS AND ARBORETA

Gardens devoted exclusively to natives

Regional Parks Botanic Garden
Intersection South Park Dr. and Wildcat Canyon Rd.,
Tilden Park, Berkeley CA 94708
(510) 841-8732
www.nativeplants.org
Open daily 8:30 A.M.–5 P.M.

Santa Barbara Botanic Garden
1212 Mission Canyon Rd., Santa Barbara CA 93105
(805) 682-4726
www.sbbg.org
Open daily 9 A.M.–6 P.M. March–October, 9 A.M.–5
P.M. November–February.

Rancho Santa Ana Botanic Garden
1500 North College Ave., Claremont CA 91711
(909) 625-8767
www.rsabg.org
Open daily 8 A.M.–5 P.M.

Gardens featuring a California natives section

UC Berkeley Botanical Garden
200 Centennial Dr., Berkeley CA 94720
(510) 642-3343
http://botanicalgarden.berkeley.edu
Open daily 9 A.M.–5 P.M.; closed first Tuesday of the
month.

UC Davis Arboretum and Botanical Gardens
Located on the UC Davis campus along a two-mile
stretch of Putah Creek. Headquarters: LaRue Rd. just
west of California Ave.
(530) 752-4880
www.arboretum.ucdavis.edu
Always open; headquarters open M–F 9 A.M.–4 P.M.

UC Santa Cruz Arboretum
1156 High St., Santa Cruz CA 95064
(831) 427-2998
www2.ucsc.edu/arboretum
Open daily 9 A.M.–5 P.M.

Quail Botanical Gardens
230 Quail Gardens Dr., Encinitas CA 92924
(760) 436-3036
www.qbgardens.com
Open daily 9 A.M.–5 P.M.

Mendocino Coast Botanical Gardens
18220 N. Hwy One, Fort Bragg CA 95437
(707) 964-4352
www.gardenbythesea.org
Open daily 9 A.M.–5 P.M. March–October, 9 A.M.–4 P.M.
November–February.

San Francisco Botanical Garden at Strybing Arboretum
9th Ave. at Lincoln Way, San Francisco CA 94122
(415) 661-1316
www.strybing.org
Open M–F 8 A.M.–4:30 P.M., weekends 10 A.M.–5 P.M.

Public gardens featuring native plants

Lindsay Wildlife Museum Nature Garden
1931 First Ave., Walnut Creek CA 94596
(925) 935-1978
www.wildlife-museum.org
Open W–F noon–5 P.M., weekends 10 A.M.–5 P.M.

Sonoma State University Campus
1801 East Cotati Ave., Rohnert Park CA 94928
(707) 664-2880
www.sonoma.edu/university

WEBSITES

www.BringingBackTheNatives.net	Offering an annual "Bringing Back the Natives" self-guided tour of East Bay gardens
www.goingnativegardentour.org	Offering an annual self-guided tour of Santa Clara Valley home gardens landscaped primarily with natives
www.calflora.net	Informative, photographic website of California natives and useful for its meanings of scientific names
www.calflora.org	CalFlora Database, a collection of data and 20,000 photos of 8,375 vascular plants of California
www.calphoto.com/wflower.htm	California Wildflower Hotsheet
www.calypteanna.com/ca-natives.htm	California Native Plants Discussion Group
www.cnplx.info	The California Native Plant Link Exchange, listing nurseries that carry natives you specify
www.gardeningwithnatives.com	Gardening group of the Santa Clara Valley Chapter of the California Native Plant Society
www.nativeplants.org	Friends of the Regional Parks Botanic Garden, Berkeley
www.rsabg.org	Rancho Santa Ana Botanic Garden
www.sbbg.org	Santa Barbara Botanic Garden
www.cnps.org	California Native Plant Society; websites of most of the thirty-three chapters can be accessed here as well.
www.gardens.com	Extensive resources for gardeners.
www.californiaoaks.org	California Oak Foundation
http://nature.berkeley.edu/comtf	California Oak Mortality Task Force

BOOKS

Bakker, Elna. *An Island Called California*. Berkeley: University of California Press, 1984. Relates plant communities, habitats, and geography in a transect across California. Highly recommended.

Bauer, Nancy. *The Habitat Garden Book: Wildlife Landscaping for the San Francisco Bay Region*. Sebastopol: Coyote Ridge Press, 2001. Good book on basic garden philosophy.

East Bay Municipal Utility District. *Plants and Landscapes for Summer-Dry Climates*. Oakland: EBMUD, 2004. Beautifully executed and profusely illustrated book on appropriate plants, designs, and gardens for the San Francisco Bay Region. Somewhat limited in its coverage of natives but otherwise an outstanding work.

Emery, Dara. *Seed Propagation of Native California Plants*. Santa Barbara: Santa Barbara Botanic Garden, 1988. Basic information on germination requirements for seeds of native plants.

Faber, Phyllis M., ed. *California's Wild Gardens: A Living Legacy*. Sacramento: California Native Plant Society, 1997. Beautifully illustrated coffee table book of special and rare habitats throughout California. Excellent color photographs.

Francis, Mark, and Andreas Reimann. *The California Landscape Garden: Ecology, Culture, and Design*. Berkeley: University of California Press, 1999. Broad-based book dealing with the reasons for cultivating natives and basic designs for using them.

Fross, David and D. Wilken. *Ceanothus*. Timber Press. 2006. Complete updated guide to the wild lilacs from a horticulture and botanical point of view.

Harlow, Nora, and Kristin Jakob, eds. *Wild Irises, Lilies, and Grasses: Gardening with California Monocots*. Berkeley: University of California Press, 2004. Beautifully illustrated book with descriptions and details of growing the most readily available, garden-worthy monocots.

Hickman, James C., ed. *The Jepson Manual*. Berkeley: University of California Press, 1993. The "bible" for identifying wild plants in California. Includes brief information on cultivation of many species.

Keator, Glenn. *Complete Garden Guide to Native Perennials*. San Francisco: Chronicle Books, 1990. A basic primer for growing a wide array of California native perennials.

Keator, Glenn. *Complete Garden Guide to Native Shrubs*. San Francisco: Chronicle Books, 1994. A basic primer for growing many California native shrubs.

Keator, Glenn. *Introduction to Trees of the San Francisco Bay Region*. Berkeley: University of California Press, 2002. Trees that occur naturally in the nine-county Bay Area.

Keator, Glenn, Linda Yamane, and Ann Lewis. *In Full View: Three Ways of Seeing California Plants*. Berkeley: Heyday Books, 1995. A botanist, artist, and Native American look at many of California's coastal plants.

Lanner, Ron. *Conifers of California*. Los Olivos: Cachuma Press, 1999. The best book on native conifers. Excellent color paintings and photographs.

Lenz, Lee. *Native Plants for California Gardens*. Claremont: Rancho Santa Ana Botanic Garden, 1956. Basic information and species for native gardens, with an emphasis on southern California.

Lenz, Lee, and John Dourley. *California Native Trees and Shrubs for Garden and Environmental Use in Southern California and Adjacent Areas*. Claremont: Rancho Santa Ana Botanic Garden, 1981. Self-explanatory title.

Lowry, Judith Larner. *Gardening with a Wild Heart: Restoring California's Native Landscapes at Home*. Berkeley: University of California Press, 1999. A compelling read on the philosophy and ways of growing California natives in a coastal landscape.

Morhardt, Sia, and Emil Morhardt. *California Desert Flowers: An Introduction to Families, Genera, and Species*. Berkeley: University of California Press, 2004. An inspiring introduction to beautiful desert plants, illustrated with excellent photographs.

Ornduff, Robert, Phyllis Faber, et al. *Introduction to California Plant Life*. Revised ed. Berkeley: University of California Press, 2003. Much basic information on California plant families, evolution, distribution, and plant communities.

Pavlik, Bruce M., Pamela C. Muick, Sharon G. Johnson, and Marjorie Popper. *Oaks of California*. Los Olivos, CA: Cachuma Press, 1993. A beautiful overview of our many species of native oaks, including identification, ecology, and preservation.

Rowntree, Lester. *Hardy Californians: A Woman's Life with Native Plants*. Expanded ed. Berkeley: University of California Press, 2006. Inspiring reading by a pioneer in natives. Highly recommended.

Rowntree, Lester. *Flowering Shrubs of California*. Currently out of print. Similar in its basic tenets to *Hardy Californians*.

Schmidt, Marjorie. *Growing California Native Plants*. Berkeley: University of California Press, 1980. One of the early basic books on growing a wide range of natives, based on personal experience.

Smith, M. Nevin. *Native Treasures: Gardening with the Plants of California*. University of California Press. 2006. Lyrical and informative descriptions and experiences growing the author's favorite native plants.

Stevens, Barbara, and Nancy Conner. *Where on Earth: A Guide to Specialty Nurseries and Other Resources for California Gardeners*. Berkeley: Heyday Books, 1999. A guide to nurseries that specialize in various kinds of plants including natives, arranged by region.

Stuart, John, and John Sawyer. *The Trees and Shrubs of California*. Berkeley: University of California Press, 2001. A good basic book for identifying native trees and a large number of native shrubs.

Wasowski, Sally, and Andy Wasowski. *Native Landscaping from El Paso to L.A.* NTC Publishing, 2000. Includes the ethic and designs of appropriate gardens incorporating natives. Somewhat skimpy on gardens in southern California, and does not cover the rest of the state.

APPENDIX 3

A CALENDAR FOR MANAGING NATIVE GARDENS

Working in the garden is a good way to connect to the earth we live on. Even though native gardens require much less work than conventional gardens, maintenance is still required. The organized gardener checks periodically to see what tasks need doing at specific times of the year. Here are some suggested jobs according to season.

Winter (December 21–March 21)

- Transplant nursery-bought plants and plants you propagated last year into the ground. Don't delay too long; late installation may mean roots are not fully established by the time summer comes along.

- Finish planting late-flowering spring bulbs in early winter.

- Finish sowing seed of spring annual and perennial wildflowers in early to midwinter.

- Finish taking hardwood cuttings for propagation by early winter.

- Finish dividing perennials, then transplant into the garden as soon as possible.

- Cut back dead growth of perennials, old flowering stalks of grasses, and fern fronds that have persisted through fall. Recycle to the compost pile.

- Assess needs for mulch and replenish as needed before plants come up in spring.

- Get a head start on the new crop of weeds, which hopefully will diminish as you mulch and plant appropriately. (Avoid rototilling or turning the soil. This action may upset the soil's balance and brings a new crop of weed seeds to the light, encouraging them to germinate. The only good reason to rototill is when soils have been heavily compacted.)

Spring (March 21–June 21)

- Keep an eye on emerging, cool-weather weeds. Weeds are easier to pull when the soil is moist after rain.

- Check on new transplants, seed pots, and cutting beds for adequate water. We often have periods of dry conditions between winter and spring rains, at which time plants are vulnerable.

- Start training shrubs and vines by tip pruning, espaliering, and pinching back. The placing of branches in training vines and espaliers is best started before new growth emerges.

- Prune out dead branches to improve air circulation in shrubs and trees during dry periods.

- As flowers appear, make notes on the timing and colors for future reference. Keeping a journal of garden events helps plan for future needs and chronicles past successes and failures.

- Catch up on labeling new starts, seed pots, and cutting beds so you'll know what you have. Many times plants will otherwise not be recognizable until they bloom. Good labeling is especially important with bulbs and grasses, which often look alike to the beginning gardener.

- Top dress potted plants in soilless mixes with a slow-release fertilizer. Unless the soil mix specifically states it has added nutrients, these fertilizers are necessary to help new cuttings and seedlings grow into vigorous plants that can later be moved into the garden.

- Make snail and slug patrols. Both can wipe out tender new growth quickly at this time of year. Avoid using chemical baits, which can have deleterious effects on the environment.

Summer (June 21–September 21)

- Remove containers with now-dormant bulbs to a cool, shaded spot until rains return. If you have beds with bulbs that are sensitive to summer water and you plan to water in summer, lift the bulbs and store them in sand in a cool, dry place until rains return.

- Potted plants need careful attention in summer. Despite how drought tolerant these plants may be, they absolutely need summer water. Be aware that plants in black plastic pots may overheat; many of these should be placed in semishaded areas to keep the roots from getting too hot. (Alternatively, use clay pots or light-colored pots.)

- Start a watering regime for plants that require summer water. These will include newly transplanted specimens (at least occasional deep watering) and plants from such communities as wetlands, riparian woodlands, and mountain meadows.

- Collect ripe seed to sow in fall or early winter. Store in envelopes or glass jars in a cool, dry place and label carefully, including the date.

- Now is the time to take tip and half-ripe wood cuttings to propagate woody perennials and small shrubs. Be sure your cutting bed is out of the direct sun and in a place where you can watch for drying out.

- Cut back long-flowering spring perennials such as California poppies to stimulate a second round of flowering and encourage new growth.

- Cut off the flowering stalks of spring-flowering perennials and compost.

- Do major pruning to shape shrubs and trees during the dry conditions of summer to prevent fungal infection. Be sure to note whether the plant you're pruning can be cut back to the ground, partway, or only tip pruned. Pruning too heavily can kill certain plants.

- Check whether beds need additional compost to keep them from drying out excessively during summer drought.

- Control warm-season weeds (weeds that germinate only after soils have warmed up) while they're young.

- Assess which natives will accept summer water, al-lowing them to stay green or bloom longer. The beauty of using natives is you often have a choice between following nature's ways or pushing the envelope a bit.

Fall (September 21–December 21)

- For plants that need even watering—plants from wetlands and riparian woodlands—be sure to continue watering. Often there is a lull between the end of summer and the start of fall rains, when plants are highly vulnerable to drying out.

- Return containers with bulbs to the garden to start growing after fall rains begin. If rains are unusually delayed, you may want to start watering the bulbs, especially if there are signs of growth.

- Prepare to transplant container plants, including vigorously rooted cuttings, as soon as soils are thoroughly wetted.

- Clean seed that you've collected. Remove as much chaff as possible to help avoid fungal infection. Seeds can be sown as soon as soils are thoroughly wetted.

- Check new cuttings to see if they need to be moved into larger containers. Generally, it's best to move cuttings into gallon cans and establish vigorous roots before moving them into the garden.

- Collect late-ripening seed. For seeds from fleshy fruits and plants from high mountains, start cold stratification in the refrigerator in anticipation of late fall or winter planting. Many plants require at least a month's stratification.

- Mark plants that go dormant to their roots in fall and winter to remind yourself where they're located.

- Gather fallen leaves you don't want to add to the compost pile. Remember, though, that most leaves make excellent mulch.

- Cut off the old growth of late-blooming perennials such as matilija poppy and hummingbird-fuchsia to encourage new growth next spring. Recycle to the compost pile.

- Dethatch your grasses and ferns by running your hands through the clumps and removing the old leaves. Use these in the compost pile. You can also use a rake or, in the case of large plants, simply cut the plants back close to the ground.

- This is the time to do some dreaming for next year's garden: peruse plant lists and catalogs and visit plant sales. New plants, of course, have to be watered until they're transplanted into the garden.

- When ordering and buying new plants, assess the garden first for the sorts of plants that will best fit in. Buying the wrong plant material can lead to mistakes that need to be corrected down the line.

APPENDIX 4

GLOSSARY OF TERMS

achene. A small, single-seeded fruit that is dispersed as a unit. Achenes do not need to have their seeds removed when planted.

alien. A nonnative plant that comes from another part of the world. Many aliens spread aggressively in their new homeland and outcompete native species.

annual. A plant that lives a year or less. Annuals need to be propagated by seed.

anthracnose. A disfiguring fungal infection in leaves that can defoliate branches.

axillary. Referring to flowers, buds, and fruits that are located in the angle between a leaf and its stem.

basalt. Dark, often black, lava-type rock.

berry. Fleshy fruit that contains several to many seeds.

biennial. A plants that lives two years. Most biennials only bloom in their second year.

bottom heat. A special cable provides heat to around 70 degrees to promote the rooting of cuttings.

bract. Modified leaf associated with flowers. Bracts may be green and look like smaller versions of leaves or may be colored and help attract pollinators. For example, the white "petals" of *Cornus nuttallii* (flowering dogwood) are actually bracts around a cluster of tiny flowers.

bunchgrass. A perennial grass that grows in dense clumps or tufts.

burl. Enlarged root crown found on some shrubs. Shrubs with burls can stump sprout when the main stems have been damaged.

capsule. A seed pod that splits into two or more sections when ripe.

catkin. Long slender spike of petalless, greenish, wind-pollinated flowers. Examples: male flowers of oaks (*Quercus* spp.) and alders (*Alnus* spp.).

compound leaf. A leaf that is divided into two or more separate leaflets. Compound leaves usually have a bud at the base of the whole leaf but not at the base of the leaflets.

cone scale. The brownish woody to papery parts of conifer seed cones that bear the seeds.

conifer. Shrubs and trees with needlelike or scalelike leaves and seeds borne in cones. Examples: pines, redwoods, cypresses, and firs.

continental climate. Climate with snowy, cold winters, short springs, and long, warm to hot, often rainy summers.

coppice. Shrubs or trees that are cut back to their root crown.

corm. Bulb-shaped underground storage organs that are solid rather than layered like the bulb of an onion.

culm. The flowering stalk of grasses.

cultivar. Cultivated variety or horticultural variety. Cultivars may stand out for having superior cold resistance, drought tolerance, longer bloom period, larger flowers, etc. Cultivar names are capitalized and placed in single quotes.

cyme. A branched flower cluster, usually flat-topped.

desert. An arid area that receives less than 10 inches of precipitation a year. Deserts may be subdivided into warm and cold deserts depending on elevation and winter temperatures.

dioecious. Species in which male and female flowers are borne on separate plants. Examples: willows (*Salix* spp.) and silk-tassel bushes (*Garrya* spp.).

disc flower. The tiny, starlike flowers in the center of a daisy or other member of the daisy family (Asteraceae).

division. Separating plants into pieces in order to propagate them. Divisions can be made from roots, underground stems, or rhizomes.

dormant. A period when plants don't grow. Dormant perennials will often die back to their roots. Many California natives are dormant in summer.

drupe. A fleshy fruit with a single or few stonelike fruits in the center. Examples: peaches and plums.

endemic. Species native and restricted to one specific geographical area.

equitant. Referring to leaves whose bases overlap in a flattened spraylike pattern, as in irises.

espalier. A shrub or woody vine that is trained to grow flat against a wall or trellis.

ferns. Perennials with usually much divided leaves that bear spore bodies underneath.

fiddlehead. When young, fern fronds are tightly coiled up and resemble the head of a fiddle.

fire-treated seed. A typical fire treatment is to place seeds in soil, add a layer of needles or other combustible material on top, and set it on fire. Care should be taken not to make the fire too hot or plant the seeds too shallowly.

floret. The tiny, petalless flower of grasses.

forest. A dense stand of trees where the tree canopies touch or overlap.

frond. The leaf, often highly divided, of a fern.

genus (pl. **genera**). A group of related species that share many characteristics in common. Examples:

pines (genus *Pinus*), oaks (genus *Quercus*), and roses (genus *Rosa*).

hardwood. The part of the stem where wood is fully developed and feels hard or tough. Some plants are best propagated from hardwood cuttings. Also applied as a general label to broadleaf flowering trees, as opposed to conifers.

head. Dense cluster of flowers at the end of a single stem.

herbaceous. Nonwoody; stems without wood or bark.

hip. The fleshy, berrylike fruit of roses, which is actually a hypanthium surrounding several hard, bony achenes.

hypanthium. A saucer- or tube-shaped structure topped by the sepals, petals, and stamens. Example: flower of the fuchsia.

indusium. A membranelike covering that protects fern sori as they develop.

layering. Propagation by laying a branch down and covering it with soil to allow rooting.

linear. Long and narrow with parallel sides.

lobed. Having leaves or other parts whose edges are indented or cut in.

Mediterranean climate. Climate with mild, rainy winters and springs, and dry, often hot summers and falls.

monoecious. Referring to species that bear male and female flowers on the same plant. Examples: oaks (*Quercus* spp.) and California hazelnut (*Corylus cornuta* var. *californica*).

native. A plant that occurs naturally in an area without the help of man. Nonnatives are brought in deliberately or by accident from other places.

nitrogen-fixing nodules. Tiny swellings on roots of the pea family (Fabaceae), alders (*Alnus* spp.), and wild lilacs (*Ceanothus* spp.) that house special bacteria that make nitrogen available in a form usable to plant roots.

node. The place on the stem where a leaf is attached.

nutlet. A small, one-seeded fruit similar to an achene.

obovate. An upside-down ovate shape.

offset. A plantlet, bulblet, or side clump of a plant that may be severed from the parent plant for vegetative propagation.

ovate. A shape that is broad and rounded near the base and gradually tapers to a pointed tip.

palmate. Arranged like the fingers on a hand; refers to veins, lobes, and leaflets of compound leaves.

panicle. The compound raceme of a flower.

perennial. A plant that lives three or more years. Perennials may be propagated by seed or vegetative means.

petals. The usually colored parts of flowers.

pinna (pl. **pinnae**). The divisions or parts of a fern frond.

pinnate. Having a featherlike pattern; applies to veins, lobes, and the arrangement of leaflets in compound leaves.

pistil. The female part of the flower, which, after fertilization, develops into a fruit with seeds.

plant community. One of many recognizable plant habitats featuring repeated dominant plants. Examples: grassland, oak woodland, redwood forest.

pollen. The often yellow dust produced by a flower's stamens. Pollen needs to be transferred to the pistil in order for seeds to form.

pollination. The process of moving pollen from the stamen to the stigma of the pistil. Pollination may be by wind, insects, birds, or other means.

pome. An applelike fruit consisting of a fleshy receptacle surrounding a papery ovary and containing several seeds. Examples: apples and pears.

prickles. Spines along stems and between leaf nodes. Example: rose stems.

raceme. Flowers borne on side branches from a long central stem.

ray (flower). The petal-like outer flowers of a daisy or other members of the daisy family (Asteraceae).

relict. A plant or animal that has continued to survive in a protected area from a time when it was more widespread.

rhizome. A sturdy, horizontal, underground stem, as found in irises, for example.

rootstock. A thickened stem at the base of a plant and often partly or completely underground.

rosette. Symmetrical cluster of leaves at the base of a plant.

samara. A winged fruit. Examples: maples (*Acer* spp.) and ashes (*Fraxinus* spp.).

scape. Leafless flowering stalk.

scarification. The process of filing away or abrading the seed coat to promote germination.

semievergreen. Plants that temporarily lose their leaves, especially in summer. Such plants will often retain their leaves with occasional summer water.

semihardwood. The same as half-ripened wood, where the stem is still somewhat bendable but is becoming firm or stiff. Some plants are best propagated from semihardwood cuttings.

sepal. The usually green, outermost layer of a flower.

serpentine. A bluish green (or red), slick, metamorphic rock that decomposes into soils low in calcium and high in heavy metals and magnesium.

serrate. Referring to leaf edges lined with sawlike teeth.

shrub. A woody plant with multiple branches or trunks.

sorus (pl. **sori**). Brown patches on the backsides of fern fronds that produce spores. The shape of the sori is useful in identifying different ferns.

species. Individual plants within a genus. Example: different kinds of oaks, such as blue oak (*Quercus douglasii*), California black oak (*Q. kelloggii*), and coast live oak (*Q. agrifolia*).

spike. A long central flower-bearing stem.

spikelets. The tiny spikes of wind-pollinated flowers of grasses and sedges.

spore. A microscopic cell produced by nonflowering plants such as ferns and horsetails. These cells grow into baby plantlets.

spur. A tapered or pointed sac that contains nectar to lure pollinators to flowers.

stamen. The male part of a flower that produces pollen.

stigma. The end of a flower's pistil, designed to receive pollen. Stigmas are generally knoblike, enlarged, or forked to provide a broad surface for pollen to cling to.

stipe. The stalk of a fern frond.

stipules. Pairs of usually small, leaflike appendages at the base of leaves. Many leaves have no stipules.

stolon. A creeping stem usually near the surface or just below the surface of the soil. Example: the runners of strawberries (*Fragaria* spp.).

stratification. A process of placing seeds in a moist medium and then chilling them in the refrigerator. Stratification may promote germination in certain kinds of seeds.

succulent. A plant with fleshy stems and/or leaves to store water against times of drought. Examples: cacti and ocotillo.

suckers. Fast-growing sprouts at the base of trees or shrubs, sometimes also developing from roots.

sweet pea-like flower. The petals are arranged this way: an upper banner, two side wings, and two fused middle petals shaped like a boat.

taproot. A long, fleshy root. Example: carrot. Taprooted plants are difficult to transplant.

tender. Sensitive to cold temperatures.

tepal. Describing flowers with sepals and petals that look alike. Example: lily (*Lilium* spp.) flowers.

thorn. A side branch that ends in a sharp spine. Example: hawthorn (*Crataegus* spp.).

transpiration. Loss of water from leaves. At least 90 percent of the water that plants take in is lost by transpiration.

tree. A woody plant with one or few large trunks.

trifoliate. Referring to compound leaves divided into three leaflets. Examples: strawberry and poison-oak leaves.

tuber. Underground, food-storing stem. Example: potato.

turf. Grasses or sedges with dense, short rhizomes that create a continuous covering in gardens.

two-lipped. An irregular flower design where the upper and lower sets of petals are separated by a petal tube. Examples: flowers of the mint family (Lamiaceae) and snapdragon family (Scrophulariaceae).

umbel. Umbrellalike cluster of flowers. Examples: onion and geranium blossoms.

variety. A variant within a species. Horticultural varieties (cultivars) may differ by only a single trait from the typical species; botanical varieties are usually founded on more substantial differences.

wash. A depression or arroyo in the landscape, generally with a higher water table than the surrounding terrain.

water shoots. Long, straight shoots that emerge from the base of trees and shrubs.

weed. An aggressive, pioneer plant that moves into disturbed habitats such as gardens, plowed fields, orchards, and roadways. Many weeds are nonnative.

whorl. A circular cluster of leaves or flowers around a stem.

woodland. Stands of trees spaced apart with openings between adjacent trees.

INDEX

Note: Bold indicates pages with photographs. Italics indicate pages with plant descriptions.